# Annals of Mathematics Studies
## Number 186

# Spaces of PL Manifolds and Categories of Simple Maps

Friedhelm Waldhausen, Bjørn Jahren and John Rognes

PRINCETON UNIVERSITY PRESS
PRINCETON AND OXFORD
2013

Published by Princeton University Press, 41 William Street,
Princeton, New Jersey 08540

In the United Kingdom: Princeton University Press, 6 Oxford Street,
Woodstock, Oxfordshire OX20 1TW

press.princeton.edu

Library of Congress Cataloging-in-Publication Data

Waldhausen, Friedhelm, 1938–
    Spaces of PL manifolds and categories of simple maps / Friedhelm Wald-
hausen, Bjørn Jahren, and John Rognes.
        pages cm. – (Annals of mathematics studies ; no. 186)
    Includes bibliographical references and index.
    ISBN 978-0-691-15775-7 (hardcover : alk. paper) – ISBN 978-0-691-15776-4
(pbk. : alk. paper) 1. Piecewise linear topology. 2. Mappings (Mathematics)
I. Jahren, Bjørn, 1945– II. Rognes, John. III. Title.
    QA613.4.W35 2013
    514'.22–dc23                                             2012038155

British Library Cataloging-in-Publication Data is available

The publisher would like to acknowledge the authors of this volume for provid-
ing the camera-ready copy from which this book was printed.

This book has been composed in $\mathcal{A}_{\mathcal{M}}\mathcal{S}$-TEX.

Printed on acid-free paper ∞

Printed in the United States of America

10 9 8 7 6 5 4 3 2 1

# Contents

Introduction                                                          1

1.  The stable parametrized $h$-cobordism theorem                     7
    1.1. The manifold part                                            7
    1.2. The non-manifold part                                       13
    1.3. Algebraic $K$-theory of spaces                              15
    1.4. Relation to other literature                                20

2.  On simple maps                                                   29
    2.1. Simple maps of simplicial sets                              29
    2.2. Normal subdivision of simplicial sets                       34
    2.3. Geometric realization and subdivision                       42
    2.4. The reduced mapping cylinder                                56
    2.5. Making simplicial sets non-singular                         68
    2.6. The approximate lifting property                            74
    2.7. Subdivision of simplicial sets over $\Delta^q$              83

3.  The non-manifold part                                            99
    3.1. Categories of simple maps                                   99
    3.2. Filling horns                                              108
    3.3. Some homotopy fiber sequences                              119
    3.4. Polyhedral realization                                     126
    3.5. Turning Serre fibrations into bundles                      131
    3.6. Quillen's Theorems A and B                                 134

4.  The manifold part                                               139
    4.1. Spaces of PL manifolds                                     139
    4.2. Spaces of thickenings                                      150
    4.3. Straightening the thickenings                              155

Bibliography                                                        175
Symbols                                                             179
Index                                                              181

Spaces of PL Manifolds and Categories of Simple Maps

# Introduction

We present a proof of the stable parametrized $h$-cobordism theorem, which we choose to state as follows:

**Theorem 0.1.** *There is a natural homotopy equivalence*

$$\mathcal{H}^{CAT}(M) \simeq \Omega \operatorname{Wh}^{CAT}(M)$$

*for each compact CAT manifold $M$, with CAT = TOP, PL or DIFF.*

Here $\mathcal{H}^{CAT}(M)$ denotes a stable CAT $h$-cobordism space defined in terms of manifolds, whereas $\operatorname{Wh}^{CAT}(M)$ denotes a CAT Whitehead space defined in terms of algebraic $K$-theory. We specify functorial models for these spaces in Definitions 1.1.1, 1.1.3, 1.3.2 and 1.3.4. See also Remark 1.3.1 for comments about our notation.

This is a stable range extension to parametrized families of the classical $h$- and $s$-cobordism theorems. Such a theorem was first stated by A. E. Hatcher in [Ha75, Thm. 9.1], but his proofs were incomplete. The aim of the present book is to provide a full proof of this key result, which provides the link between the geometric topology of high-dimensional manifolds and their automorphisms, and the algebraic $K$-theory of spaces and structured ring spectra. The book is based on a manuscript by the first author, with the same title, which was referred to as "to appear" in [Wa82] and as "to appear (since '79)" in [Wa87b].

We first recall the classical $h$- and $s$-cobordism theorems. Let $M$ be a compact manifold of dimension $d$, either in the topological, piecewise-linear or smooth differentiable sense. An $h$-cobordism on $M$ is a compact $(d+1)$-manifold $W$ whose boundary decomposes as a union $\partial W \cong M \cup N$ of two codimension zero submanifolds along their common boundary, such that each inclusion $M \subset W$ and $N \subset W$ is a homotopy equivalence. Two $h$-cobordisms $W$ and $W'$ on $M$ are isomorphic if there is a homeomorphism, PL homeomorphism or diffeomorphism $W \cong W'$, as appropriate for the geometric category, that restricts to the identity on $M$. An $h$-cobordism is said to be trivial if it is isomorphic to the product $h$-cobordism $W = M \times [0,1]$, containing $M$ as $M \times 0$. (There is a little technical point here about corners in the DIFF case, which we gloss over.)

Assume for simplicity that $M$ is connected and has a chosen base point, let $\pi = \pi_1(M)$ be its fundamental group, and let $\operatorname{Wh}_1(\pi) = K_1(\mathbb{Z}[\pi])/(\pm\pi)$ be the Whitehead group of $\pi$, which is often denoted $\operatorname{Wh}(\pi)$. To each $h$-cobordism $W$ on $M$ there is associated an element $\tau(W, M) \in \operatorname{Wh}_1(\pi)$, called its Whitehead torsion [Mi66, §9]. Trivial $h$-cobordisms have zero torsion. In general, an $h$-cobordism with zero torsion is called an $s$-cobordism. The $h$-cobordism theorem of S. Smale (concerning the simply-connected case, when the Whitehead group is trivial) and the $s$-cobordism theorem of D. Barden, B. Mazur and J. R. Stallings (for arbitrary fundamental groups), assert for $d \geq 5$ that the Whitehead torsion defines a one-to-one correspondence

$$\{h\text{-cobordisms on } M\}/(\text{iso}) \xrightarrow{\cong} \mathrm{Wh}_1(\pi)$$

$$[W] \mapsto \tau(W, M)$$

between the isomorphism classes of $h$-cobordisms on $M$ and the elements of the Whitehead group. Thus, in these dimensions the $s$-cobordisms are precisely the trivial $h$-cobordisms.

This result should be viewed as the computation of the set of path components of the **space of $h$-cobordisms**, and the aim of a **parametrized** $h$-cobordism theorem is to determine the homotopy type of this space. More precisely, for $M$ a CAT manifold there is a space $H^{CAT}(M)$ that classifies CAT bundles of $h$-cobordisms on $M$, and the $s$-cobordism theorem asserts (for $d \geq 5$) that there is a natural bijection

$$\pi_0 H^{CAT}(M) \cong \mathrm{Wh}_1(\pi).$$

We shall have to settle for a **stable** parametrized $h$-cobordism theorem, which provides a homotopy equivalent model for the stabilized $h$-cobordism space

$$\mathcal{H}^{CAT}(M) = \operatorname*{colim}_{k} H^{CAT}(M \times [0,1]^k).$$

The model in question is defined in algebraic $K$-theoretic terms, much like the definition of the Whitehead group by means of the algebraic $K_1$-group of the integral group ring $\mathbb{Z}[\pi]$. We will express each CAT Whitehead space $\mathrm{Wh}^{CAT}(M)$ in terms of the algebraic $K$-theory space $A(M)$, which was introduced (largely for this purpose) by the first author. As stated at the outset, the model for the homotopy type of $\mathcal{H}^{CAT}(M)$ will be the based loop space $\Omega\,\mathrm{Wh}^{CAT}(M)$. The PL and TOP Whitehead spaces $\mathrm{Wh}^{PL}(M) = \mathrm{Wh}^{TOP}(M)$ will be the same, because it is known by triangulation theory that $H^{PL}(M) \simeq H^{TOP}(M)$ for PL manifolds of dimension $d \geq 5$, but $\mathrm{Wh}^{DIFF}(M)$ will be different. (This wordplay is due to Hatcher.)

By definition, the **algebraic $K$-theory $A(M)$ of the space $M$** is the loop space $\Omega|hS_\bullet\mathcal{R}_f(M)|$ of the geometric realization of the subcategory of homotopy equivalences in the $S_\bullet$-construction on the category with cofibrations and weak equivalences $\mathcal{R}_f(M)$ of finite retractive spaces over $M$. See [Wa85, §2.1]. Iteration of the $S_\bullet$-construction specifies a preferred sequence of higher deloopings of $A(M)$, so we may view that space as the underlying infinite loop space of a spectrum $\mathbf{A}(M)$, in the sense of algebraic topology.

Letting $*$ denote a one-point space, the PL Whitehead space is defined so that there is a natural homotopy fiber sequence

$$h(M; A(*)) \xrightarrow{\alpha} A(M) \to \mathrm{Wh}^{PL}(M)$$

for each space $M$, where $h(M; A(*)) = \Omega^\infty(\mathbf{A}(*) \wedge M_+)$ is the unreduced generalized homology of $M$ with coefficients in the spectrum $\mathbf{A}(*)$, and $\alpha$ is the natural assembly map to the homotopy functor $A(M)$. The stable parametrized PL $h$-cobordism theorem can therefore be restated as follows.

**Theorem 0.2.** *There is a natural homotopy fiber sequence*

$$\mathcal{H}^{PL}(M) \to h(M; A(*)) \xrightarrow{\alpha} A(M)$$

*for each compact PL manifold $M$, where $\alpha$ is the assembly map.*

The DIFF Whitehead space is defined so that there is a natural homotopy fiber sequence

$$Q(M_+) \xrightarrow{\iota} A(M) \to \mathrm{Wh}^{DIFF}(M)$$

for each space $M$, where $Q(M_+) = \Omega^\infty \Sigma^\infty(M_+)$ and $\iota$ is induced by the unit map $\eta \colon \mathbf{S} \to \mathbf{A}(*)$ and the assembly map. In this case the fiber sequence is naturally split up to homotopy, by a stabilization map $A(M) \to A^S(M) \simeq Q(M_+)$. The stable parametrized DIFF $h$-cobordism theorem can therefore be restated as follows.

**Theorem 0.3.** *There is a natural homotopy fiber sequence*

$$\mathcal{H}^{DIFF}(M) \to Q(M_+) \xrightarrow{\iota} A(M)$$

*for each compact DIFF manifold $M$. The map $\iota$ is naturally split, up to homotopy, so there is a natural homotopy factorization*

$$A(M) \xrightarrow{\simeq} Q(M_+) \times \mathrm{Wh}^{DIFF}(M) \,,$$

*where $\mathrm{Wh}^{DIFF}(M) \simeq B\mathcal{H}^{DIFF}(M)$.*

We note that when $M$ is based and connected, with fundamental group $\pi$, then $\pi_1 A(M) \cong K_1(\mathbb{Z}[\pi])$ and $\pi_1 \mathrm{Wh}^{CAT}(M) \cong \mathrm{Wh}_1(\pi)$, so these theorems recover the classical identification of $\pi_0 H^{CAT}(M)$ with $\mathrm{Wh}_1(\pi)$, but only after stabilization.

There is a third appearance of the Whitehead group functor, in addition to those arising from manifold topology and algebraic $K$-theory, given in terms of general topological spaces in [St70]. Given a compact polyhedron $K$ we can consider the class of compact polyhedra $L$ that contain $K$ as a deformation retract, and say that two such polyhedra $L$ and $L'$ are equivalent if they can be connected by a finite chain of elementary deformations (expansions and collapses) relative to $K$. The set of equivalence classes is the Whitehead group $\mathrm{Wh}_1(K)$, which for connected $K$ with fundamental group $\pi$ is isomorphic to $\mathrm{Wh}_1(\pi)$. Alternatively, we may form a category $s\mathcal{E}^h(K)$, with objects the compact polyhedra $L$ containing $K$ such that the inclusion $K \subset L$ is a homotopy equivalence, and morphisms $L \to L'$ the PL maps relative to $K$ that have contractible point inverses. The Whitehead group $\mathrm{Wh}_1(K)$ is then the set of path components of $s\mathcal{E}^h(K)$, and the homotopy types of several closely related categories play a central role in this book.

The proof of the PL case of Theorem 0.1, as presented in this work, can be summarized in the following diagram of simplicial sets, categories and simplicial categories.

$$
(0.4)
$$

$$
\begin{array}{ccccc}
\mathcal{H}^{PL}(M) & \longrightarrow & \operatorname{colim}_n \mathcal{M}^n_\bullet & \xrightarrow{\ j\ } & \operatorname{colim}_n h\mathcal{M}^n_\bullet \\
\downarrow{\scriptstyle u} & & \downarrow{\scriptstyle u} & & \downarrow{\scriptstyle u} \\
s\widetilde{\mathcal{E}}^h_\bullet(M) & \longrightarrow & s\widetilde{\mathcal{E}}_\bullet & \xrightarrow{\ j\ } & h\widetilde{\mathcal{E}}_\bullet \\
\uparrow{\scriptstyle \tilde{n}r} & & \uparrow{\scriptstyle \tilde{n}r} & & \uparrow{\scriptstyle \tilde{n}r} \\
s\mathcal{D}^h(X) & \longrightarrow & s\mathcal{D} & \xrightarrow{\ j\ } & h\mathcal{D} \\
\downarrow{\scriptstyle i} & & \downarrow{\scriptstyle i} & & \downarrow{\scriptstyle i} \\
s\mathcal{C}^h(X) & \longrightarrow & s\mathcal{C} & \xrightarrow{\ j\ } & h\mathcal{C}
\end{array}
$$

Here $X$ is a finite combinatorial manifold, with geometric realization the compact PL manifold $M = |X|$. The notation $\mathcal{M}^n_\bullet$ refers to PL bundles of stably framed compact $n$-manifolds, the notation $\widetilde{\mathcal{E}}_\bullet$ refers to PL Serre fibrations of compact polyhedra, the notation $\mathcal{D}$ refers to finite non-singular simplicial sets (a little more general than ordered simplicial complexes) and the notation $\mathcal{C}$ refers to general finite simplicial sets. The prefixes $s$ and $h$ refer to categories of simple maps (with contractible point inverses) and (weak) homotopy equivalences, respectively. The entries in the left hand column refer to objects which, in a suitable sense, contain $M$ or $X$ as a deformation retract.

We show in Sections 3.2, 3.3 and 4.1 that the horizontal rows are homotopy fiber sequences. In Sections 4.1, 4.2 and 4.3 we show that the middle and right hand vertical maps $u$ are homotopy equivalences. These maps view stably framed PL manifolds as polyhedra, and PL bundles as PL Serre fibrations. In Sections 3.4 and 3.5 we show that the middle and right hand vertical maps $\tilde{n}r$ are homotopy equivalences. These pass from simplicial complexes to polyhedra by geometric realization, and simultaneously introduce a simplicial direction by viewing product bundles as PL Serre fibrations. In Section 3.1 we show that the middle and right hand vertical maps $i$ are homotopy equivalences. These view non-singular simplicial sets as general simplicial sets. Chapter 2 contains the foundational material on simple maps of simplicial sets needed for these proofs. More detailed explanations and references are given in Sections 1.1 and 1.2.

Taken together, these results show that we have natural homotopy equivalences

$$
\mathcal{H}^{PL}(M) \simeq s\widetilde{\mathcal{E}}^h_\bullet(M) \simeq s\mathcal{D}^h(X) \simeq s\mathcal{C}^h(X) \,.
$$

The nerve of the category $s\mathcal{C}^h(X)$ is one of several naturally equivalent models for the looped PL Whitehead space $\Omega \operatorname{Wh}^{PL}(X)$, by Proposition 3.1.1, Theorem 3.1.7 and Theorem 3.3.1 of [Wa85]:

$$
s\mathcal{C}^h(X) \simeq \Omega|s\mathcal{N}_\bullet\mathcal{C}^h(X)| \simeq \Omega|sS_\bullet\mathcal{R}^h_f(X^{\Delta^\bullet})| \simeq \Omega \operatorname{Wh}^{PL}(X) \,.
$$

Its relation to the functor $A(X)$, leading to Theorems 0.2 and 0.3 and the DIFF case of Theorem 0.1, is reviewed in Section 1.3.

At the level of path components, the first chain of homotopy equivalences gives the classification of (stable) $h$-cobordisms in terms of the topologically defined Whitehead group, see Corollary 3.2.4, while the second chain of homotopy equivalences makes the connection to the algebraically defined Whitehead group.

In Section 1.4, we comment on the relation of this book (and a paper of K. Igusa [Ig88]) to Hatcher's paper [Ha75], reaching the conclusion that in spite of the missing technical details of the latter, its main conclusions turn out to be qualitatively correct. We also discuss an alternative approach to the PL case of Theorem 0.1, which with some effort can be assembled from papers by M. Steinberger [St86] and T. A. Chapman [Ch87], in combination with [Wa85] and parts of the present book.

Chapter 1 has been written to present only the definitions needed to precisely state the main results. Chapter 2 contains all the technical material on simple maps of simplicial sets and related constructions. Some of the details are quite intricate, and it may be more enjoyable to go through them after learning why these results are useful. The geometrically minded reader, interested in $h$-cobordisms and spaces of PL manifolds, could start with Chapter 4 instead. The algebraic $K$-theorist, interested in $A(X)$ and categories of simple maps, could start with Chapter 3. Both of the latter two chapters have been written with this possibility in mind, and contain direct references to Chapter 2 for most of the technical results needed.

A previous version of this work was developed around 1990 by the first author and Wolrad Vogell. It improved on the ordering of the material from the original manuscript, and added to the exposition of the manifold part. However, some years later that development had stalled. The second and third author joined the project around the year 2000, and are both grateful for being given the opportunity to help complete this important bridge between high-dimensional geometric topology and the algebraic $K$-theory of spaces.

We are grateful to Tom Goodwillie for an explanation of the PL concordance stability theorem for smoothable manifolds (Corollary 1.4.2), to Larry Siebenmann for a clarification of the meaning of zero Whitehead torsion for pairs of simplicial complexes (Lemma 3.2.9(d)), and to Dan Grayson for advice on the exposition. Thanks are also due to the referee for constructive comments and questions, as well as to Johannes Ebert, Philipp Kühl, Martin Olbermann and Henrik Rüping of the Graduiertenkolleg Bonn–Düsseldorf–Bochum seminar for their comments.

# Chapter One

## The stable parametrized *h*-cobordism theorem

### 1.1. THE MANIFOLD PART

We write DIFF for the category of $C^\infty$ smooth manifolds, PL for the category of piecewise-linear manifolds, and TOP for the category of topological manifolds. We generically write CAT for any one of these geometric categories. Let $I = [0, 1]$ and $J$ be two fixed closed intervals in $\mathbb{R}$. We will form collars using $I$ and stabilize manifolds and polyhedra using $J$.

In this section, as well as in Chapter 4, we let $\Delta^q = \{(t_0, \ldots, t_q) \mid \sum_{i=0}^q t_i = 1, t_i \geq 0\}$ be the standard affine $q$-simplex.

By a **CAT bundle** $\pi \colon E \to \Delta^q$ we mean a CAT locally trivial family, i.e., a map such that there exists an open cover $\{U_\alpha\}$ of $\Delta^q$ and a CAT isomorphism over $U_\alpha$ (= a local trivialization) from $\pi^{-1}(U_\alpha) \to U_\alpha$ to a product bundle, for each $\alpha$. For $\pi$ to be a CAT bundle **relative to** a given product subbundle, we also ask that each local trivialization restricts to the identity on the product subbundle. We can always shrink the open cover to a cover by compact subsets $\{K_\alpha\}$, whose interiors still cover $\Delta^q$, and this allows us to only work with compact polyhedra in the PL case.

**Definition 1.1.1.** (a) Let $M$ be a compact CAT manifold, with empty or nonempty boundary. We define the **CAT $h$-cobordism space** $H(M) = H^{CAT}(M)$ of $M$ as a simplicial set. Its 0-simplices are the compact CAT manifolds $W$ that are $h$-cobordisms on $M$, i.e., the boundary

$$\partial W = M \cup N$$

is a union of two codimension zero submanifolds along their common boundary $\partial M = \partial N$, and the inclusions

$$M \subset W \supset N$$

are homotopy equivalences. For each $q \geq 0$, a $q$-simplex of $H(M)$ is a CAT bundle $\pi \colon E \to \Delta^q$ relative to the trivial subbundle $pr \colon M \times \Delta^q \to \Delta^q$, such that each fiber $W_p = \pi^{-1}(p)$ is a CAT $h$-cobordism on $M \cong M \times p$, for $p \in \Delta^q$.

(b) We also define a **collared** CAT $h$-cobordism space $H(M)^c = H^{CAT}(M)^c$, whose 0-simplices are $h$-cobordisms $W$ on $M$ equipped with a choice of collar, i.e., a CAT embedding

$$c \colon M \times I \to W$$

that identifies $M \times 0$ with $M$ in the standard way, and takes $M \times [0, 1)$ to an open neighborhood of $M$ in $W$. A $q$-simplex of $H(M)^c$ is a CAT bundle

$\pi\colon E \to \Delta^q$ relative to an embedded subbundle $pr\colon M \times I \times \Delta^q \to \Delta^q$, such that each fiber is a collared CAT $h$-cobordism on $M$. The map $H(M)^c \to H(M)$ that forgets the choice of collar is a weak homotopy equivalence, because spaces of collars are contractible.

*Remark 1.1.2.* To ensure that these collections of simplices are really sets, we might assume that each bundle $E \to \Delta^q$ is embedded in $\mathbb{R}^\infty \times \Delta^q \to \Delta^q$. The simplicial operator associated to $\alpha\colon \Delta^p \to \Delta^q$ takes $E \to \Delta^q$ to the image of the pullback $\alpha^*(E) \subset \Delta^p \times_{\Delta^q} (\mathbb{R}^\infty \times \Delta^q)$ under the canonical identification $\Delta^p \times_{\Delta^q} (\mathbb{R}^\infty \times \Delta^q) \cong \mathbb{R}^\infty \times \Delta^p$. See [HTW90, 2.1] for a more detailed solution. To smooth any corners that arise, we interpret DIFF manifolds as coming equipped with a smooth normal field, as in [Wa82, §6]. The emphasis in this book will be on the PL case.

To see that the space of CAT collars on $M$ in $W$ is contractible, we note that [Ar70, Thm. 2] proves that any two TOP collars are ambient isotopic (relative to the boundary), and the argument generalizes word-for-word to show that any two parametrized families of collars (over the same base) are connected by a family of ambient isotopies, which proves the claim for TOP. In the PL category, the same proof works, once PL isotopies are chosen to replace the TOP isotopies $F_s$ and $G_s$ given on page 124 of [Ar70]. The proof in the DIFF case is different, using the convexity of the space of inward pointing normal fields.

**Definition 1.1.3.** (a) The **stabilization map**

$$\sigma\colon H(M) \to H(M \times J)$$

takes an $h$-cobordism $W$ on $M$ to the $h$-cobordism $W \times J$ on $M \times J$. It is well-defined, because $M \times J \subset W \times J$ and $(N \times J) \cup (W \times \partial J) \subset W \times J$ are homotopy equivalences. The **stable $h$-cobordism space** of $M$ is the colimit

$$\mathcal{H}^{CAT}(M) = \operatorname*{colim}_k H^{CAT}(M \times J^k)$$

over $k \geq 0$, formed with respect to the stabilization maps. Each stabilization map is a cofibration of simplicial sets, so the colimit has the same homotopy type as the corresponding homotopy colimit, or mapping telescope.

(b) In the collared case, the stabilization map $\sigma\colon H(M)^c \to H(M \times J)^c$ takes a collared $h$-cobordism $(W, c)$ on $M$ to the $h$-cobordism $W \times J$ on $M \times J$ with collar

$$M \times I \times J \xrightarrow{c \times id} W \times J.$$

Each codimension zero CAT embedding $M \to M'$ induces a map $H(M)^c \to H(M')^c$ that takes $(W, c)$ to the $h$-cobordism

$$W' = M' \times I \cup_{M \times I} W,$$

with the obvious collar $c'\colon M' \times I \to W'$. This makes $H(M)^c$ and $\mathcal{H}^{CAT}(M)^c = \operatorname{colim}_k H(M \times J^k)^c$ covariant **functors** in $M$, for codimension zero embeddings of CAT manifolds. The forgetful map $\mathcal{H}^{CAT}(M)^c \to \mathcal{H}^{CAT}(M)$ is also a weak homotopy equivalence.

We must work with the collared $h$-cobordism space when functoriality is required, but will often (for simplicity) just refer to the plain $h$-cobordism space. To extend the functoriality from codimension zero embeddings to general continuous maps $M \to M'$ of topological spaces, one can proceed as in [Ha78, Prop. 1.3] or [Wa82, p. 152], to which we refer for details.

*Remark 1.1.4.* For a cobordism to become an $h$-cobordism after suitable stabilization, weaker homotopical hypotheses suffice. For example, let $X \subset V$ be a codimension zero inclusion and homotopy equivalence of compact CAT manifolds. Let $c_0\colon \partial X \times I \to X$ be an interior collar on the boundary of $X$, let $M_0 = c_0(\partial X \times 1)$ and $W_0 = c_0(\partial X \times I) \cup (V \setminus X)$. Then $W_0$ is a cobordism from $M_0$ to $N_0 = \partial V$, and the inclusion $M_0 \subset W_0$ is a homology equivalence by excision, but $W_0$ is in general not an $h$-cobordism on $M_0$. However, if we stabilize the inclusion $X \subset V$ three times, and perform the corresponding constructions, then the resulting cobordism is an $h$-cobordism.

In more detail, we have a codimension zero inclusion and homotopy equivalence $X \times J^3 \subset V \times J^3$. Choosing an interior collar $c\colon \partial(X \times J^3) \times I \to X \times J^3$ on the boundary of $X \times J^3$, we let $M = c(\partial(X \times J^3) \times 1)$, $N = \partial(V \times J^3)$ and

$$ W = c(\partial(X \times J^3) \times I) \cup (V \times J^3 \setminus X \times J^3) . $$

Then $W$ is a cobordism from $M$ to $N$. The three inclusions $M \subset X \times J^3$, $N \subset V \times J^3$ and $W \subset V \times J^3$ are all $\pi_1$-isomorphisms (because any null-homotopy in $V \times J^3$ of a loop in $N$ can be deformed away from the interior of $V$ times some interior point of $J^3$, and then into $N$, and similarly in the two other cases). Since $X \times J^3 \subset V \times J^3$ is a homotopy equivalence, it follows that both $M \subset W$ and $N \subset W$ are $\pi_1$-isomorphisms. By excision, it follows that $M \subset W$ is a homology equivalence, now with arbitrary local coefficients. By the universal coefficient theorem, and Lefschetz duality for the compact manifold $W$, it follows that $N \subset W$ is a homology equivalence, again with arbitrary local coefficients. Hence both $M \subset W$ and $N \subset W$ are homotopy equivalences, and $W$ is an $h$-cobordism on $M$.

In the following definitions, we specify one model $s\widetilde{\mathcal{E}}^h_\bullet(M)$ for the stable PL $h$-cobordism space $\mathcal{H}^{PL}(M)$, based on a category of compact polyhedra and simple maps. In the next two sections we will re-express this polyhedral model: first in terms of a category of finite simplicial sets and simple maps, and then in terms of the algebraic $K$-theory of spaces.

**Definition 1.1.5.** A PL map $f\colon K \to L$ of compact polyhedra will be called a **simple map** if it has contractible point inverses, i.e., if $f^{-1}(p)$ is contractible for each point $p \in L$. (A space is contractible if it is homotopy equivalent to a one-point space. It is, in particular, then non-empty.)

In this context, M. Cohen [Co67, Thm. 11.1] has proved that simple maps (which he called contractible mappings) are simple homotopy equivalences.

Two compact polyhedra are thus of the same simple homotopy type if and only if they can be linked by a finite chain of simple maps. The composite of two simple maps is always a simple map. This follows from Proposition 2.1.3 in Chapter 2, in view of the possibility of triangulating polyhedra and PL maps. Thus we can interpret the simple homotopy types of compact polyhedra as the path components of (the nerve of) a category of polyhedra and simple maps.

**Definition 1.1.6.** Let $K$ be a compact polyhedron. We define a simplicial category $s\widetilde{\mathcal{E}}_\bullet^h(K)$ of compact polyhedra containing $K$ as a deformation retract, and simple PL maps between these. In simplicial degree 0, the objects are compact polyhedra $L$ equipped with a PL embedding and homotopy equivalence $K \to L$. The morphisms $f\colon L \to L'$ are the simple PL maps that restrict to the identity on $K$, via the given embeddings. A deformation retraction $L \to K$ exists for each object, but a choice of such a map is not part of the structure.

In simplicial degree $q$, the objects of $s\widetilde{\mathcal{E}}_q^h(K)$ are **PL Serre fibrations** ($=$ PL maps whose underlying continuous map of topological spaces is a Serre fibration) of compact polyhedra $\pi\colon E \to \Delta^q$, with a PL embedding and homotopy equivalence $K \times \Delta^q \to E$ over $\Delta^q$ from the product fibration $pr\colon K \times \Delta^q \to \Delta^q$. The morphisms $f\colon E \to E'$ of $s\widetilde{\mathcal{E}}_q^h(K)$ are the simple PL fiber maps over $\Delta^q$ that restrict to the identity on $K \times \Delta^q$, via the given embeddings.

Each PL embedding $K \to K'$ of compact polyhedra induces a (forward) functor $s\widetilde{\mathcal{E}}_\bullet^h(K) \to s\widetilde{\mathcal{E}}_\bullet^h(K')$ that takes $K \to L$ to $K' \to K' \cup_K L$, and similarly in parametrized families. The pushout $K' \cup_K L$ exists as a polyhedron, because both $K \to K'$ and $K \to L$ are PL embeddings. This makes $s\widetilde{\mathcal{E}}_\bullet^h(K)$ a covariant functor in $K$, for PL embeddings. There is a natural **stabilization map**

$$\sigma\colon s\widetilde{\mathcal{E}}_\bullet^h(K) \to s\widetilde{\mathcal{E}}_\bullet^h(K \times J)$$

that takes $K \to L$ to $K \times J \to L \times J$, and similarly in parametrized families. It is a homotopy equivalence by Lemma 4.1.12 in Chapter 4.

As in the following definition, we often regard a simplicial set as a simplicial category with only identity morphisms, a simplicial category as the bisimplicial set given by its degreewise nerve (Definition 2.2.1), and a bisimplicial set as the simplicial set given by its diagonal. A map of categories, i.e., a functor, is a homotopy equivalence if the induced map of nerves is a weak homotopy equivalence. See [Se68, §2], [Qu73, §1] or [Wa78a, §5] for more on these conventions.

**Definition 1.1.7.** Let $M$ be a compact PL manifold. There is a natural map of simplicial categories

$$u\colon H^{PL}(M)^c \to s\widetilde{\mathcal{E}}_\bullet^h(M \times I)$$

that takes $(W, c)$ to the underlying compact polyhedron of the $h$-cobordism $W$, with the PL embedding and homotopy equivalence provided by the collar $c\colon M \times I \to W$, and views PL bundles over $\Delta^q$ as being particular cases of

PL Serre fibrations over $\Delta^q$. It commutes with the stabilization maps, and therefore induces a natural map

$$u \colon \mathcal{H}^{PL}(M)^c \to \operatorname*{colim}_k s\widetilde{\mathcal{E}}^h_{\bullet}(M \times I \times J^k) \,.$$

Here is the PL manifold part of the stable parametrized $h$-cobordism theorem.

**Theorem 1.1.8.** *Let $M$ be a compact PL manifold. There is a natural homotopy equivalence*

$$\mathcal{H}^{PL}(M) \simeq s\widetilde{\mathcal{E}}^h_{\bullet}(M) \,.$$

*More precisely, there is a natural chain of homotopy equivalences*

$$\mathcal{H}^{PL}(M)^c = \operatorname*{colim}_k H^{PL}(M \times J^k)^c \xrightarrow[\simeq]{u} \operatorname*{colim}_k s\widetilde{\mathcal{E}}^h_{\bullet}(M \times I \times J^k) \xleftarrow[\simeq]{\sigma} s\widetilde{\mathcal{E}}^h_{\bullet}(M) \,,$$

*and $\mathcal{H}^{PL}(M)^c \simeq \mathcal{H}^{PL}(M)$.*

By the argument of [Wa82, p. 175], which we explain below, it suffices to prove Theorem 1.1.8 when $M$ is a codimension zero submanifold of Euclidean space, or a little more generally, when $M$ is stably framed (see Definition 4.1.2). The proof of the stably framed case will be given in Chapter 4, and is outlined in Section 4.1. Cf. diagram (4.1.13).

*Remark 1.1.9 (Reduction of Theorem 1.1.8 to the stably framed case).* Here we use a second homotopy equivalent model $H(M)^r$ for the $h$-cobordism space of $M$, where each $h$-cobordism $W$ comes equipped with a choice of a CAT retraction $r \colon W \to M$, and similarly in parametrized families. The forgetful map $H(M)^r \to H(M)$ is a weak homotopy equivalence, because each inclusion $M \subset W$ is a cofibration and a homotopy equivalence. For each CAT disc bundle $\nu \colon N \to M$ there is a **pullback map** $\nu^! \colon H(M)^r \to H(N)^r$, which takes an $h$-cobordism $W$ on $M$ with retraction $r \colon W \to M$ to the pulled-back $h$-cobordism $N \times_M W$ on $N$, with the pulled-back retraction.

$$
\begin{array}{ccccc}
M \times J^k & \overset{\simeq}{\rightarrowtail} & W \times J^k & \xrightarrow{r \times id} & M \times J^k \\
\tau \downarrow & & \downarrow & & \downarrow \tau \\
N & \overset{\simeq}{\rightarrowtail} & N \times_M W & \longrightarrow & N \\
\nu \downarrow & & \downarrow r^* \nu & & \downarrow \nu \\
M & \overset{\simeq}{\rightarrowtail} & W & \xrightarrow{r} & M
\end{array}
$$

If $\tau \colon M \times J^k \to N$ is a second CAT disc bundle, so that the composite $\nu\tau$ equals the projection $pr \colon M \times J^k \to M$, then $(\nu\tau)^!$ equals the $k$-fold stabilization map $\tau^!\nu^! = \sigma^k$. Hence there is a commutative diagram

$$
\begin{array}{ccc}
H(M)^r & \xrightarrow{\nu^!} H(N)^r \xrightarrow{\tau^!} & H(M \times J^k)^r \\
\simeq \downarrow & & \downarrow \simeq \\
H(M) & \xrightarrow{\quad \sigma^k \quad} & H(M \times J^k) \,.
\end{array}
$$

According to Haefliger–Wall [HW65, Cor. 4.2], each compact PL manifold $M$ admits a stable normal disc bundle $\nu\colon N \to M$, with $N$ embedded with codimension zero in some Euclidean $n$-space. Furthermore, PL disc bundles admit stable inverses. Let $\tau\colon M \times J^k \to N$ be the disc bundle in such a stable inverse to $\nu$, such that $\nu\tau$ is isomorphic to the product $k$-disc bundle over $M$, and $\tau(\nu \times id)$ is isomorphic to the product $k$-disc bundle over $N$. By the diagram above, pullback along $\nu$ and $\tau$ define homotopy inverse maps

$$\mathcal{H}^{PL}(M) \xrightarrow{\nu^!} \mathcal{H}^{PL}(N) \xrightarrow{\tau^!} \mathcal{H}^{PL}(M)$$

after stabilization.

Likewise, there is a homotopy equivalent variant $s\widetilde{\mathcal{E}}^h_\bullet(M)^r$ of $s\widetilde{\mathcal{E}}^h_\bullet(M)$, with a (contractible) choice of PL retraction $r\colon L \to M$ for each polyhedron $L$ containing $M$, and similarly in parametrized families. There is a simplicial functor $\nu^!\colon s\widetilde{\mathcal{E}}^h_\bullet(M)^r \to s\widetilde{\mathcal{E}}^h_\bullet(N)^r$, by the pullback property of simple maps (see Proposition 2.1.3). It is a homotopy equivalence, because each stabilization map $\sigma$ is a homotopy equivalence by Lemma 4.1.12. Thus it suffices to prove Theorem 1.1.8 for $N$, which is stably framed, in place of $M$.

*Remark 1.1.10.* A similar argument lets us reduce the stable parametrized TOP $h$-cobordism theorem to the PL case. By [Mi64] and [Ki64] each compact TOP manifold $M$ admits a normal disc bundle $\nu\colon N \to M$ in some Euclidean space, and $\nu$ admits a stable inverse. As a codimension zero submanifold of Euclidean space, $N$ can be given a PL structure. By the argument above, $\nu^!\colon \mathcal{H}^{TOP}(M) \to \mathcal{H}^{TOP}(N)$ is a homotopy equivalence. Furthermore, $H^{PL}(N) \to H^{TOP}(N)$ is a homotopy equivalence for $n = \dim(N) \geq 5$, by triangulation theory [BL74, Thm. 6.2] and [KS77, V.5.5]. Thus $\mathcal{H}^{PL}(N) \simeq \mathcal{H}^{TOP}(N)$, and the TOP case of Theorem 0.1 follows from the PL case.

*Remark 1.1.11.* There are further possible variations in the definition of the $h$-cobordism space $H(M)$. For a fixed $h$-cobordism $W$ on $M$, the path component of $H(M)$ containing $W$ is a classifying space for CAT bundles with fiber $W$, relative to the product bundle with fiber $M$. A homotopy equivalent model for this classifying space is the bar construction $BCAT(W \operatorname{rel} M)$ of the simplicial group of CAT automorphisms of $W$ relative to $M$. Hence there is a homotopy equivalence

$$H(M) \simeq \coprod_{[W]} BCAT(W \operatorname{rel} M)\,,$$

where $[W]$ ranges over the set of isomorphism classes of CAT $h$-cobordisms on $M$.

In particular, when $W = M \times I$ is the product $h$-cobordism on $M \cong M \times 0$, we are led to the simplicial group

$$(1.1.12) \qquad\qquad C(M) = CAT(M \times I, M \times 1)$$

of **CAT concordances** (= pseudo-isotopies) on $M$. By definition, these are the CAT automorphisms of $M \times I$ that pointwise fix the complement of $M \times 1$

in $\partial(M \times I)$. More generally, we follow the convention of [WW01, 1.1.2] and write $CAT(W, N)$ for the simplicial group of CAT automorphisms of $W$ that agree with the identity on the complement of $N$ in $\partial W$. Here $N$ is assumed to be a codimension zero CAT submanifold of the boundary $\partial W$. When $N$ is empty we may omit it from the notation, so that $CAT(W) = CAT(W \operatorname{rel} \partial W)$.

The concordances that commute with the projection to $I = [0, 1]$ are the same as the isotopies of $M \operatorname{rel} \partial M$ that start from the identity, but concordances are not required to commute with this projection, hence the name pseudo-isotopy. The inclusion $C(M) \to CAT(M \times I \operatorname{rel} M \times 0)$ is a homotopy equivalence, so the path component of $H(M)$ that contains the trivial $h$-cobordisms is homotopy equivalent to the bar construction $BC(M)$. In general, $H(M)$ is a non-connective delooping of the CAT concordance space $C(M)$.

By the $s$-cobordism theorem, the set of path components of $H(M)$ is in bijection with the Whitehead group $\operatorname{Wh}_1(\pi) = K_1(\mathbb{Z}[\pi])/(\pm\pi)$, when $d = \dim(M) \geq 5$ and $M$ is connected with fundamental group $\pi$. For disconnected $M$, the Whitehead group should be interpreted as the sum of the Whitehead groups associated to its individual path components. For each element $\tau \in \operatorname{Wh}_1(\pi)$, we write $H(M)_\tau$ for the path component of $H(M)$ that consists of the $h$-cobordisms with Whitehead torsion $\tau$. For example, $H(M)_0 \simeq BC(M)$ is the $s$-cobordism space.

Still assuming $d \geq 5$, we can find an $h$-cobordism $W_1$ from $M$ to $M_\tau$, with prescribed Whitehead torsion $\tau$ relative to $M$, and a second $h$-cobordism $W_2$ from $M_\tau$ to $M$, with Whitehead torsion $-\tau$ relative to $M_\tau$. Then $W_1 \cup_{M_\tau} W_2 \cong M \times I$ and $W_2 \cup_M W_1 \cong M_\tau \times I$, by the sum formula for Whitehead torsion and the $s$-cobordism theorem. Gluing with $W_2$ at $M$, and with $W_1$ at $M_\tau$, define homotopy inverse maps

$$H(M)_\tau \to H(M_\tau)_0 \to H(M)_\tau \, .$$

Hence

$$(1.1.13) \qquad H(M) = \coprod_\tau H(M)_\tau \simeq \coprod_\tau H(M_\tau)_0 \simeq \coprod_\tau BC(M_\tau) \, ,$$

where $\tau \in \operatorname{Wh}_1(\pi)$.

## 1.2. The non-manifold part

In this section, as well as in Chapters 2 and 3, we let $\Delta^q$ be the simplicial $q$-simplex, the simplicial set with geometric realization $|\Delta^q|$ the standard affine $q$-simplex.

**Definition 1.2.1.** A simplicial set $X$ is **finite** if it is generated by finitely many simplices, or equivalently, if its geometric realization $|X|$ is compact. A map $f \colon X \to Y$ of finite simplicial sets will be called a **simple map** if its geometric realization $|f| \colon |X| \to |Y|$ has contractible point inverses, i.e., if for each $p \in |Y|$ the preimage $|f|^{-1}(p)$ is contractible.

A map $f\colon X \to Y$ of simplicial sets is a **weak homotopy equivalence** if its geometric realization $|f|$ is a homotopy equivalence. A map $f\colon X \to Y$ of simplicial sets is a **cofibration** if it is injective in each degree, or equivalently, if its geometric realization $|f|$ is an embedding. We say that $f$ is a **finite cofibration** if, furthermore, $Y$ is generated by the image of $X$ and finitely many other simplices.

We shall see in Section 2.1 that simple maps are weak homotopy equivalences, and that the composite of two simple maps is a simple map. In particular, the simple maps of finite simplicial sets form a category.

**Definition 1.2.2.** By the Yoneda lemma, there is a one-to-one correspondence between the $n$-simplices $x$ of a simplicial set $X$ and the simplicial maps $\bar{x}\colon \Delta^n \to X$. We call $\bar{x}$ the **representing map** of $x$. A simplicial set $X$ will be called **non-singular** if for each non-degenerate simplex $x \in X$ the representing map $\bar{x}\colon \Delta^n \to X$ is a cofibration.

In any simplicial set $X$, the geometric realization $|\bar{x}|\colon |\Delta^n| \to |X|$ of the representing map of a non-degenerate simplex $x$ restricts to an embedding of the interior of $|\Delta^n|$. The additional condition imposed for non-singular simplicial sets is that this map is required to be an embedding of the whole of $|\Delta^n|$. It amounts to the same to ask that the images of the $(n+1)$ vertices of $|\Delta^n|$ in $|X|$ are all distinct.

When viewed as simplicial sets, ordered simplicial complexes provide examples of non-singular simplicial sets, but not all non-singular simplicial sets arise this way. For example, the union $\Delta^1 \cup_{\partial\Delta^1} \Delta^1$ of two 1-simplices along their boundary is a non-singular simplicial set, but not an ordered simplicial complex.

**Definition 1.2.3.** For any simplicial set $X$, let $\mathcal{C}(X)$ be the category of finite cofibrations $y\colon X \to Y$. The morphisms from $y$ to $y'\colon X \to Y'$ are the simplicial maps $f\colon Y \to Y'$ under $X$, i.e., those satisfying $fy = y'$.

For finite $X$, let $s\mathcal{C}^h(X) \subset \mathcal{C}(X)$ be the subcategory with objects such that $y\colon X \to Y$ is a weak homotopy equivalence, and morphisms such that $f\colon Y \to Y'$ is a simple map. Let $\mathcal{D}(X) \subset \mathcal{C}(X)$ and $s\mathcal{D}^h(X) \subset s\mathcal{C}^h(X)$ be the full subcategories generated by the objects $y\colon X \to Y$ for which $Y$ is non-singular. Let $i\colon s\mathcal{D}^h(X) \to s\mathcal{C}^h(X)$ be the inclusion functor.

The definition of $s\mathcal{C}^h(X)$ only makes sense, as stated, for finite $X$, because we have not defined what it means for $f\colon Y \to Y'$ to be a simple map when $Y$ or $Y'$ are not finite. We will extend the definition of $s\mathcal{C}^h(X)$ to general simplicial sets $X$ in Definition 3.1.12, as the colimit of the categories $s\mathcal{C}^h(X_\alpha)$ where $X_\alpha$ ranges over the finite simplicial subsets of $X$. The categories $\mathcal{D}(X)$ and $s\mathcal{D}^h(X)$ are only non-empty when $X$ itself is non-singular, because there can only be a cofibration $y\colon X \to Y$ to a non-singular simplicial set $Y$ when $X$ is also non-singular.

**Definition 1.2.4.** The geometric realization $|X|$ of a finite non-singular simplicial set $X$ is canonically a compact polyhedron, which we call the **polyhedral**

**realization** of $X$. Its polyhedral structure is characterized by the condition that $|\bar{x}| \colon |\Delta^n| \to |X|$ is a PL map for each (non-degenerate) simplex $x$ of $X$. The geometric realization $|f| \colon |X| \to |Y|$ of a simplicial map of finite non-singular simplicial sets is then a PL map.

For any compact polyhedron $K$, let $s\mathcal{E}^h(K)$ be the category of PL embeddings $\ell \colon K \to L$ of compact polyhedra, and simple PL maps $f \colon L \to L'$ under $K$. For any finite non-singular simplicial set $X$ let $r \colon s\mathcal{D}^h(X) \to s\mathcal{E}^h(|X|)$ be the polyhedral realization functor that takes $y \colon X \to Y$ to $|y| \colon |X| \to |Y|$, and similarly for morphisms. Let $\tilde{n} \colon s\mathcal{E}^h(K) \to s\widetilde{\mathcal{E}}^h_\bullet(K)$ be the simplicial functor that includes $s\mathcal{E}^h(K)$ as the 0-simplices in $s\widetilde{\mathcal{E}}^h_\bullet(K)$, as introduced in Definition 1.1.6.

See Definition 3.4.1 for more on compact polyhedra, PL maps and the polyhedral realization functor. The non-manifold parts of the stable parametrized $h$-cobordism theorem follow.

**Theorem 1.2.5.** *Let $X$ be a finite non-singular simplicial set. The full inclusion functor*

$$i \colon s\mathcal{D}^h(X) \to s\mathcal{C}^h(X)$$

*is a homotopy equivalence.*

Theorem 1.2.5 will be proved as part of Proposition 3.1.14. Cf. diagram (3.1.15).

**Theorem 1.2.6.** *Let $X$ be a finite non-singular simplicial set. The composite*

$$\tilde{n} \circ r \colon s\mathcal{D}^h(X) \to s\widetilde{\mathcal{E}}^h_\bullet(|X|)$$

*of the polyhedral realization functor $r$ and the 0-simplex inclusion $\tilde{n}$, is a homotopy equivalence.*

Theorem 1.2.6 is proved at the end of Section 3.5. Cf. diagram (3.5.4). We do not claim that the individual functors $r \colon s\mathcal{D}^h(X) \to s\mathcal{E}^h(|X|)$ and $\tilde{n} \colon s\mathcal{E}^h(|X|) \to s\widetilde{\mathcal{E}}^h_\bullet(|X|)$ are homotopy equivalences, only their composite. The proof involves factoring the composite in a different way, through a simplicial category $s\widetilde{\mathcal{D}}^h_\bullet(X)$, to be introduced in Definition 3.1.7(d).

The construction $X \mapsto s\mathcal{C}^h(X)$ is covariantly functorial in the simplicial set $X$. It is homotopy invariant in the sense that any weak homotopy equivalence $x \colon X \to X'$ induces a homotopy equivalence $x_* \colon s\mathcal{C}^h(X) \to s\mathcal{C}^h(X')$. Union along $X$ defines a sum operation on $s\mathcal{C}^h(X)$ that makes it a grouplike monoid, with $\pi_0 s\mathcal{C}^h(X)$ isomorphic to the Whitehead group of $\pi_1(X)$. See Definition 3.1.11, Corollary 3.2.4 and Proposition 3.2.5 for precise statements and proofs.

## 1.3. ALGEBRAIC $K$-THEORY OF SPACES

For any simplicial set $X$, let $\mathcal{R}_f(X)$ be the category of finite retractive spaces over $X$, with objects $(Y, r, y)$ where $y \colon X \to Y$ is a finite cofibration of simplicial sets and $r\colon Y \to X$ is a retraction, so that $ry = id_X$. A morphism from $(Y, r, y)$ to $(Y', r', y')$ is a simplicial map $f\colon Y \to Y'$ over and under $X$, so that $r = r'f$ and $fy = y'$. There is a functor $\mathcal{R}_f(X) \to \mathcal{C}(X)$ that forgets the structural retractions. (The category $\mathcal{C}(X)$ was denoted $\mathcal{C}_f(X)$ in [Wa78b] and [Wa85], but in this book we omit the subscript to make room for a simplicial direction.)

The two subcategories $co\mathcal{R}_f(X)$ and $h\mathcal{R}_f(X)$ of $\mathcal{R}_f(X)$, of maps $f\colon Y \to Y'$ that are cofibrations and weak homotopy equivalences, respectively, make $\mathcal{R}_f(X)$ a category with cofibrations and weak equivalences in the sense of [Wa85, §1.1 and §1.2]. The $S_\bullet$-construction $S_\bullet \mathcal{R}_f(X)$ is then defined as a simplicial category (with cofibrations and weak equivalences), see [Wa85, §1.3], and the **algebraic $K$-theory** of the space $X$ is defined to be the loop space

$$A(X) = \Omega |hS_\bullet \mathcal{R}_f(X)| .$$

Any weak homotopy equivalence $X \to X'$ induces a homotopy equivalence $A(X) \to A(X')$, and we can write $A(M)$ for $A(X)$ when $M = |X|$.

The $S_\bullet$-construction can be iterated, and the sequence of spaces

$$\{\, n \mapsto |h\underbrace{S_\bullet \cdots S_\bullet}_{n} \mathcal{R}_f(X)| \,\}$$

(with appropriate structure maps) defines a spectrum $\mathbf{A}(X)$, which has $A(X)$ as its underlying infinite loop space. Let $\mathbf{S} = \{n \mapsto S^n\}$ be the **sphere spectrum**. In the special case $X = *$ there is a unit map

$$\eta \colon \mathbf{S} \to \mathbf{A}(*) ,$$

adjoint to the based map $S^0 \to |h\mathcal{R}_f(*)|$ that takes the non-base point to the 0-simplex corresponding to the object $(Y, r, y)$ with $Y = S^0$.

These spectra can be given more structure. By [GH99, Prop. 6.1.1] each $\mathbf{A}(X)$ is naturally a symmetric spectrum [HSS00], with the symmetric group $\Sigma_n$ acting on the $n$-th space by permuting the $S_\bullet$-constructions. Furthermore, the smash product of finite based simplicial sets induces a multiplication $\mu \colon \mathbf{A}(*) \wedge \mathbf{A}(*) \to \mathbf{A}(*)$ that, together with the unit map $\eta$, makes $\mathbf{A}(*)$ a commutative symmetric ring spectrum. Each spectrum $\mathbf{A}(X)$ is naturally an $\mathbf{A}(*)$-module spectrum.

For based and connected $X$, there is a homotopy equivalent definition of $A(X)$ as the algebraic $K$-theory $K(\mathbf{S}[\Omega X])$ of the spherical group ring $\mathbf{S}[\Omega X]$. Here $\Omega X$ can be interpreted as the Kan loop group of $X$, see [Wa96], and $\mathbf{S}[\Omega X]$ is its unreduced suspension spectrum $\Sigma^\infty(\Omega X)_+$, viewed as a symmetric ring spectrum, or any other equivalent notion.

*Remark 1.3.1.* The CAT Whitehead spaces can be defined in several, mostly equivalent, ways. In early papers on the subject [Wa78b, pp. 46–47], [Wa82, p. 144], [WW88, pp. 575–576], $\text{Wh}^{CAT}(M)$ is defined for compact CAT manifolds $M$ as a delooping of the stable $h$-cobordism space $\mathcal{H}^{CAT}(M)$, making

Theorem 0.1 a definition rather than a theorem. With that definition in mind, the reader might justifiably wonder what this book is all about.

On the other hand, in [Ha75, p. 102] and [Ha78, p. 15] the PL Whitehead space $\mathrm{Wh}^{PL}(K)$ is defined for polyhedra $K$ as a delooping of the classifying space of the category of simple maps that we denote by $s\mathcal{E}^h(K)$. (In Hatcher's first cited paper, there is no delooping.) In [Wa85, Prop. 3.1.1] the PL Whitehead space $\mathrm{Wh}^{PL}(X)$ is defined for simplicial sets $X$ as the delooping $|s\mathcal{N}_\bullet \mathcal{C}^h(X)|$ of the classifying space of the category $s\mathcal{C}^h(X)$. We do not know that Hatcher and Waldhausen's definitions are equivalent for $K = |X|$, but they do become equivalent if $s\mathcal{E}^h(K)$ is expanded to the simplicial category $s\widetilde{\mathcal{E}}^h_\bullet(K)$, see diagram (3.1.8) and Remark 3.1.10.

With Waldhausen's cited definition, the PL case of Theorem 0.1 becomes the main result established in this book, asserting that there is a natural equivalence $\mathcal{H}^{PL}(|X|) \simeq s\mathcal{C}^h(X)$ for finite combinatorial manifolds $X$, by the proof outlined in diagram (0.4). This definition has the advantage that it provides notation for stating Theorem 0.1 in the PL case, but it has the disadvantage that it does not also cover the DIFF case.

To obtain the given statement of Theorem 0.1, and to directly connect the main result about $h$-cobordism spaces to algebraic $K$-theory, we therefore choose to redefine the CAT Whitehead spaces $\mathrm{Wh}^{CAT}(X)$ directly in terms of the functor $A(X)$, by analogy with the definition of the Whitehead group $\mathrm{Wh}_1(\pi)$ as a quotient of the algebraic $K$-group $K_1(\mathbb{Z}[\pi])$. The role of the geometric category CAT is not apparent in the resulting definition of $\mathrm{Wh}^{CAT}(X)$, so the superscript in the notation is only justified once Theorem 0.1 has been proved.

That the $K$-theoretic definition in the PL case agrees with Waldhausen's cited definition is the content of [Wa85, Thm. 3.1.7] and [Wa85, Thm. 3.3.1]. The correctness of the redefinition in the DIFF case (which is the real content of the DIFF case of Theorem 0.1) is a consequence of smoothing theory and a vanishing theorem, and is explained at the end of this section.

By [Wa85, Thm. 3.2.1] and a part of [Wa85, Thm. 3.3.1] (recalled in diagram (1.4.7) below), there is a natural map

$$\alpha \colon h(X; A(*)) \to A(X)$$

of homotopy functors in $X$, where

$$h(X; A(*)) = \Omega^\infty(\mathbf{A}(*) \wedge X_+)$$

is the unreduced homological functor associated to the spectrum $\mathbf{A}(*)$. The natural map $\alpha$ is a homotopy equivalence for $X = *$, which characterizes it up to homotopy equivalence as the **assembly map** associated to the homotopy functor $A(X)$, see [WW95, §1]. The assembly map extends to a map

$$\alpha \colon \mathbf{A}(*) \wedge X_+ \to \mathbf{A}(X)$$

of (symmetric) spectra, as is seen from [Wa85, Thm. 3.3.1] by iterating the $S_\bullet$-construction.

**Definition 1.3.2.** For each simplicial set $X$, let the **PL Whitehead spectrum** $\mathbf{Wh}^{PL}(X)$ be defined as the homotopy cofiber of the spectrum level assembly map, so that there is a natural cofiber sequence of spectra

$$\mathbf{A}(*) \wedge X_+ \xrightarrow{\alpha} \mathbf{A}(X) \to \mathbf{Wh}^{PL}(X).$$

Let the **PL Whitehead space** $\mathrm{Wh}^{PL}(X)$ be defined as the underlying infinite loop space $\mathrm{Wh}^{PL}(X) = \Omega^\infty \mathbf{Wh}^{PL}(X)$.

Let the **TOP Whitehead spectrum** and **TOP Whitehead space** be defined in the same way, as $\mathbf{Wh}^{TOP}(X) = \mathbf{Wh}^{PL}(X)$ and $\mathrm{Wh}^{TOP}(X) = \mathrm{Wh}^{PL}(X)$, respectively.

With this (revised) definition, there is obviously a natural homotopy fiber sequence

(1.3.3) $$h(X; A(*)) \xrightarrow{\alpha} A(X) \to \mathrm{Wh}^{PL}(X)$$

of homotopy functors in $X$. Continuing the homotopy fiber sequence one step to the left, we get an identification of the looped PL Whitehead space $\Omega\,\mathrm{Wh}^{PL}(X)$ with the homotopy fiber of the space level assembly map $\alpha\colon h(X; A(*)) \to A(X)$, without needing to refer to the previously mentioned spectrum level constructions.

*Summary of proof of the PL case of Theorem 0.1, and Theorem 0.2.* By [Wa85, Thm. 3.1.7] and [Wa85, 3.3.1], the revised definition of $\mathrm{Wh}^{PL}(X)$ agrees up to natural homotopy equivalence with the one given in [Wa85, Prop. 3.1.1]. In particular, there is a natural chain of homotopy equivalences

$$s\mathcal{C}^h(X) \simeq \Omega\,\mathrm{Wh}^{PL}(X),$$

also with the revised definition. The proof of [Wa85, Thm. 3.1.7] contains some forward references to results proved in the present book, which we have summarized in Remark 1.4.5.

By our Theorems 1.1.8, 1.2.5 and 1.2.6, proved in Sections 4.1–4.3, 3.1 and 3.5, respectively, there is a natural chain of homotopy equivalences

$$\mathcal{H}^{PL}(M) \xrightarrow{\simeq} s\tilde{\mathcal{E}}^h_\bullet(M) \xleftarrow{\simeq} s\mathcal{D}^h(X) \xrightarrow{\simeq} s\mathcal{C}^h(X)$$

for each compact PL manifold $M$, triangulated as $|X|$. This establishes the homotopy equivalence of Theorem 0.1 in the PL case. The homotopy fiber sequence of Theorem 0.2 is the Puppe sequence obtained by continuing (1.3.3) one step to the left. $\square$

The unit map $\eta\colon \mathbf{S} \to \mathbf{A}(*)$ induces a natural map of unreduced homological functors

$$Q(X_+) = \Omega^\infty(\mathbf{S} \wedge X_+) \xrightarrow{\eta} \Omega^\infty(\mathbf{A}(*) \wedge X_+) = h(X; A(*)).$$

We define the spectrum map $\iota\colon \Sigma^\infty X_+ \to \mathbf{A}(X)$ as the composite

$$\Sigma^\infty X_+ = \mathbf{S} \wedge X_+ \xrightarrow{\eta \wedge id} \mathbf{A}(*) \wedge X_+ \xrightarrow{\alpha} \mathbf{A}(X)$$

and let $\iota = \alpha \circ \eta\colon Q(X_+) \to A(X)$ be the underlying map of infinite loop spaces.

**Definition 1.3.4.** For each simplicial set $X$ let the **DIFF Whitehead spectrum** $\mathbf{Wh}^{DIFF}(X)$ be defined as the homotopy cofiber of the spectrum map $\iota$, so that there is a natural cofiber sequence of spectra

$$\Sigma^\infty X_+ \xrightarrow{\iota} \mathbf{A}(X) \to \mathbf{Wh}^{DIFF}(X)\,.$$

Let the **DIFF Whitehead space** $\mathrm{Wh}^{DIFF}(X)$ be defined as the underlying infinite loop space $\mathrm{Wh}^{DIFF}(X) = \Omega^\infty \mathbf{Wh}^{DIFF}(X)$.

There is obviously a natural homotopy fiber sequence

(1.3.5)                     $Q(X_+) \xrightarrow{\iota} A(X) \to \mathrm{Wh}^{DIFF}(X)$

of homotopy functors in $X$. Continuing the homotopy fiber sequence one step to the left, we get an identification of the looped DIFF Whitehead space $\Omega\,\mathrm{Wh}^{DIFF}(X)$ with the homotopy fiber of the space level map $\iota\colon Q(X_+) \to A(X)$. However, in this case the splitting of $\iota$ leads to the attractive formula $A(X) \simeq Q(X_+) \times \mathrm{Wh}^{DIFF}(X)$, which is one reason to focus on the unlooped Whitehead space.

*Proof of the DIFF case of Theorem 0.1, and Theorem 0.3.* We can deduce Theorem 0.3 and the DIFF case of Theorem 0.1 from Theorem 0.2. The argument was explained in [Wa78b, §3] and [Wa82, §2], but we review and comment on it here for the reader's convenience.

We consider homotopy functors $F$ from spaces to based spaces, such that there is a natural map $F(M) \to \mathrm{hofib}(F(M_+) \to F(*))$. The **stabilization** $F^S$ of $F$ (not related to the other kind of stabilization that we use) is an unreduced homological functor, with

$$F^S(M) \simeq \operatorname*{colim}_n \Omega^n \,\mathrm{hofib}(F(\Sigma^n(M_+)) \to F(*))\,.$$

In the notation of [Go90b], $F^S(M) = D_*F(M_+)$, where $D_*F$ is the differential of $F$ at $*$. There is a natural map $F(M) \to F^S(M)$, which is a homotopy equivalence whenever $F$ itself is a homological functor. This form of stabilization preserves natural homotopy fiber sequences.

Each term in the homotopy fiber sequence of Theorem 0.2 is such a homotopy functor. Hence there is a natural homotopy equivalence

$$\Omega\,\mathrm{hofib}(A(M) \to A^S(M)) \xrightarrow{\simeq} \mathrm{hofib}(\mathcal{H}^{PL}(M) \to \mathcal{H}^{PL,S}(M))\,.$$

The stable $h$-cobordism space $\mathcal{H}^{DIFF}(M)$ can also be extended to such a homotopy functor. By Morlet's disjunction lemma [BLR75, §1], cf. [Ha78, Lem. 5.4], the stabilized functor $\mathcal{H}^{DIFF,S}(M)$ is contractible. By smoothing theory, also known as Morlet's comparison theorem, the homotopy fiber of the natural map $\mathcal{H}^{DIFF}(M) \to \mathcal{H}^{PL}(M)$ is a homological functor [BL77, §4]. Hence there is a natural chain of homotopy equivalences

$$\mathcal{H}^{DIFF}(M) \xleftarrow{\simeq} \mathrm{hofib}(\mathcal{H}^{DIFF}(M) \to \mathcal{H}^{DIFF,S}(M))$$
$$\xrightarrow{\simeq} \mathrm{hofib}(\mathcal{H}^{PL}(M) \to \mathcal{H}^{PL,S}(M))\,.$$

The composite map $Q(M_+) \xrightarrow{\iota} A(M) \to A^S(M)$ is a homotopy equivalence, by the "vanishing of the mystery homology theory" [Wa87a, Thm.]. Alternatively, this can be deduced from B. I. Dundas' theorem on relative $K$-theory [Du97, p. 224], which implies that the cyclotomic trace map induces a profinite homotopy equivalence $A^S(M) \simeq TC^S(M)$, together with the calculation $TC^S(M) \simeq Q(M_+)$ of [He94]. The rational result was obtained in [Wa78b, Prop. 2.9] from work by A. Borel [Bo74], F. T. Farrell and W.-C. Hsiang [FH78]. Either way, it follows that the composite natural map

$$\mathrm{hofib}(A(M) \to A^S(M)) \to A(M) \to \mathrm{Wh}^{DIFF}(M)$$

is a homotopy equivalence. In combination, we obtain a natural chain of homotopy equivalences that induces the homotopy equivalence

$$\mathcal{H}^{DIFF}(M) \simeq \Omega\,\mathrm{Wh}^{DIFF}(M)$$

claimed in Theorem 0.1. The homotopy fiber sequence of Theorem 0.3 is the Puppe sequence obtained by continuing (1.3.5) one step to the left. The stabilization map $A(M) \to A^S(M)$ provides a natural splitting of $\iota\colon Q(M_+) \to A(M)$, up to homotopy, and together with the map $A(M) \to \mathrm{Wh}^{DIFF}(M)$ it defines the natural homotopy factorization of the theorem. $\quad\square$

## 1.4. RELATION TO OTHER LITERATURE

The main assertion in Hatcher's paper [Ha75] is his Theorem 9.1, saying that there is a $k$-connected map from the PL $h$-cobordism space $H^{PL}(M)$ to a classifying space $\mathcal{S}(M)$ for "PL Serre fibrations with homotopy fiber $M$ and a fiber homotopy trivialization," provided that $n = \dim(M) \geq 3k + 5$. The model for $\mathcal{S}(M)$ chosen by Hatcher equals the simplicial set of objects in our simplicial category $s\widetilde{\mathcal{E}}^h_\bullet(M)$. In Hatcher's Proposition 3.1, this space is asserted to be homotopy equivalent to the nerve of $s\mathcal{E}^h(M)$. That particular claim appears to be difficult to prove in the polyhedral context, because the proposed argument for his Proposition 2.5 makes significant use of chosen triangulations. However, it follows from [St86, Thm. 1] and our Theorem 1.2.6 that $\mathcal{S}(M)$ is homotopy equivalent to the nerve of the simplicial category $s\widetilde{\mathcal{E}}^h_\bullet(M)$, so in essence, Hatcher's Theorem 9.1 claims that the map $H^{PL}(M) \to s\widetilde{\mathcal{E}}^h_\bullet(M)$ is about $(n/3)$-connected, for $n = \dim(M)$. Stabilizing with respect to the dimension, this amounts to the manifold part Theorem 1.1.8 of our stable parametrized $h$-cobordism theorem. Thus the stable form of Hatcher's main assertion is correct.

The relevance of simple maps to the study of PL homeomorphisms of manifolds may be motivated by the following theorem of M. Cohen [Co70, Thm. 1]: For closed PL $n$-manifolds $M$ and $N$ with $n \geq 5$ each simple PL map $M \to N$ can be uniformly approximated by a PL homeomorphism $M \cong N$. A similar result in the TOP category was proved by L. Siebenmann [Si72].

The first author's paper [Wa78b] (from the 1976 Stanford conference) contains in its Section 5 the assertion that Hatcher's polyhedral model $s\widetilde{\mathcal{E}}_\bullet^h(M)$ for $\mathcal{H}^{PL}(M)$ is homotopy equivalent to the model $s\mathcal{C}^h(X)$ that is defined in terms of simplicial sets, where $M = |X|$ as usual. This translation is the content of our non-manifold Theorems 1.2.5 and 1.2.6. Furthermore, Section 5 of that paper contains the homotopy fiber sequences of Theorems 0.2 and 0.3. Modulo some forward references to the present work, their proofs appeared in [Wa85], except for the result that $A^S(M) \simeq Q(M_+)$, which appeared in [Wa87a]. For more on these forward references, see Remark 1.4.5.

Hatcher's paper [Ha78] in the same proceedings surveys, among other things, how concordance spaces (with their canonical involution) measure the difference between the "honest" automorphism groups of manifolds and the block automorphism groups of manifolds, which are determined by surgery theory [Wa70, §17.A]. The spectral sequence of [Ha78, Prop. 2.1] makes this precise in the concordance stable range. In [WW88, Thm. A], M. Weiss and B. Williams express this spectral sequence as coming from the $\mathbb{Z}/2$-homotopy orbit spectral sequence of an involution on the stable $h$-cobordism space, with its infinite loop space structure. Their later survey [WW01] explains, among many other things, how this contribution from concordance and $h$-cobordism spaces also measures the difference between the "honest" moduli space parametrizing bundles of compact manifolds and the block moduli space given by the surgery classification of manifolds.

In the meantime, M. Steinberger's paper [St86] appeared, whose Theorem 1 proves that (the nerve of) $s\mathcal{D}^h(X)$ is a classifying space for "PL Serre fibrations with homotopy fiber $|X|$ and a fiber homotopy trivialization." Thus $s\mathcal{D}^h(X) \simeq \mathcal{S}(M)$, which is close to our Theorem 1.2.6. His main tool for proving this is a special category of finite convex cell complexes in Euclidean space, and certain piecewise linear maps between these.

Steinberger's Theorem 2 is the same as our Theorem 1.2.5, but his proof leaves a significant part to be discovered by the reader. His argument [St86, p. 19] starts out just as our first (non-functorial) proof of Proposition 3.1.14, and relies on a result similar to our Proposition 2.5.1. At that point, he appeals to an analog $C(h)$ of Cohen's PL mapping cylinder, but defined for general maps $h$ of simplicial sets. However, he does not establish the existence of this construction, nor its relevant properties. Presumably the intended $C(h)$ is our backward reduced mapping cylinder $M(Sd(h))$ of the normal subdivision of $h$, and the required properties are those established in our Sections 2.1 through 2.4.

The following year, T. A. Chapman's paper [Ch87] appeared. His Theorem 3 proves the stable form of Hatcher's main claim, that a version of $\mathcal{H}^{PL}(M)$ is homotopy equivalent to the classifying space $\mathcal{S}(M)$. Modulo the identification of $\mathcal{S}(M)$ with $s\widetilde{\mathcal{E}}_\bullet^h(M)$, this is equivalent to our Theorem 1.1.8. Combining Chapman's Theorem 3 with Steinberger's Theorems 1 and 2 one obtains a homotopy equivalence $\mathcal{H}^{PL}(M) \simeq s\mathcal{C}^h(X)$, for $M = |X|$. When combined with the homotopy equivalence $s\mathcal{C}^h(X) \simeq \Omega \mathrm{Wh}^{PL}(X)$ from [Wa85, §3], bringing algebraic $K$-theory into the picture, one recovers the PL case of our Theorem 0.1. In a similar way, Chapman's Theorem 2 is analogous to our main geometric Theorem 4.1.14, except that Chapman works with manifolds embedded with codimension zero in some Euclidean space, whereas we have chosen to work with stably framed manifolds. His main tool is a stable fibered controlled $h$-cobordism theorem.

Chapman's paper omits proofs of several results, because of their similarity with other results in the literature (his Propositions 2.2 and 2.3), and only discusses the absolute case of some inductive proofs that rely on a relative statement for their inductive hypotheses (his Theorems 3.2 and 5.2). Furthermore, some arguments involving careful control estimates are only explained over the 0- and 1-skeleta of a parameter domain, and it is left to the reader to extend these over all higher skeleta.

Since Theorem 1.1.8, 1.2.5 and 1.2.6 are fundamental results for the relation between the stable $h$-cobordism spaces and the Whitehead spaces, we prefer to provide proofs that do not leave too many constructions, generalizations or relativizations to be discovered or filled in by the reader. The tools used in our presentation are close to those of [Wa85], which provides the connection onwards from the Whitehead spaces to the algebraic $K$-theory of spaces. Taken together, these two works complete the bridge connecting geometric topology to algebraic $K$-theory.

The present book is also needed to justify the forward references from [Wa85], including Theorem 2.3.2 and its consequence Proposition 2.3.3, which were used in [Wa85, §3.1] on the way to Theorem 0.2. Hence these results from our Chapter 2 are also required for Theorem 0.3 and the DIFF case of Theorem 0.1, neither of which are covered by Steinberger and Chapman's papers.

Returning to Hatcher's original paper, the unstable form of the main assertion would imply not only the stable conclusion, but also a PL concordance stability result [Ha78, Cor. 9.2], to the effect that a suspension map $\sigma\colon C^{PL}(M) \to C^{PL}(M \times J)$ is about $(n/3)$-connected, for $n = \dim(M)$. Delooping once, this would imply that the stabilization map $\sigma\colon H^{PL}(M) \to H^{PL}(M \times J)$ is also about $(n/3)$-connected. As we discuss in Remark 4.2.3, our methods are essentially stable. In particular, we do not attempt to prove these PL concordance stability results. However, working in the DIFF category, K. Igusa proved the following concordance stability result in [Ig88], using Hatcher's PL argument as an outline for the proof.

**Theorem 1.4.1 (Igusa).** *The suspension map*

$$\sigma\colon C^{DIFF}(M) \to C^{DIFF}(M \times J)$$

*is $k$-connected, for all compact smooth $n$-manifolds $M$ with $n \geq \max\{2k + 7, 3k + 4\}$.*

Delooping once, and iterating, it follows that the infinite stabilization map $H^{DIFF}(M) \to \mathcal{H}^{DIFF}(M)$ is $(k + 1)$-connected, for $M$, $n$ and $k$ as in the theorem. When combined with Theorem 0.3 and calculations of the algebraic $K$-theory of spaces $A(M)$, this leads to concrete results on the homotopy groups $\pi_i C^{DIFF}(M)$ and $\pi_i H^{DIFF}(M)$, for $i$ up to about $n/3$.

For example, in the case $M = D^n \simeq *$ there is a rational homotopy equivalence $A(M) \simeq A(*) \to K(\mathbb{Z})$, and the striking consequences for $\pi_i DIFF(D^n) \otimes \mathbb{Q}$ of Borel's calculation [Bo74] of $K_i(\mathbb{Z}) \otimes \mathbb{Q}$ were explained in [Wa78b, Thm. 3.2] and [Ig88, p. 7]. Recall from Remark 1.1.11 the convention that $DIFF(D^n) = DIFF(D^n \operatorname{rel} S^{n-1})$. Analogous rational results for Euclidean and spherical space forms were obtained in [FH78], [HJ82] and [HJ83]. Calculations of the $p$-torsion in $\pi_i A(*)$ were made in [Ro02] for $p = 2$ and [Ro03] for odd regular primes, and some consequences concerning the $p$-torsion in $\pi_i DIFF(D^n)$ were drawn in Section 6 of the latter paper.

D. Burghelea and R. Lashof [BL77, Thm. C] used smoothing theory and Morlet's disjunction lemma to show that the PL concordance stability theorem stated by Hatcher would imply a DIFF concordance stability theorem, in about half the PL concordance stable range. T. Goodwillie has improved on this argument, using his multiple disjunction lemma from [Go90a], to establish a DIFF concordance stable range only three less than such an assumed PL concordance stable range.

However, no proof of a concordance stability theorem for general PL manifolds seems to be known. In the absence of a PL proof, it was observed by Burghelea and by Goodwillie that for smoothable manifolds $M$ one can deduce a PL concordance stability theorem from Igusa's DIFF concordance stability theorem, with the same concordance stable range. The following argument was explained to us by Goodwillie. It implies that the optimal DIFF concordance stable range and the optimal PL concordance stable range for smoothable manifolds are practically the same.

**Corollary 1.4.2 (Burghelea, Goodwillie).** *The suspension map*

$$\sigma \colon C^{PL}(M) \to C^{PL}(M \times J)$$

*is $k$-connected, for compact smoothable $n$-manifolds $M$ with $n \geq \max\{2k + 7, 3k + 4\}$.*

*Proof.* Let $M$ be a compact DIFF $n$-manifold and let $P \to M$ be its frame bundle, i.e., the principal $O_n$-bundle associated to the tangent bundle of $M$. By smoothing theory [BL74, Thm. 4.2] there is a homotopy fiber sequence

$$DIFF(M) \to PL(M) \to \Gamma(M; PL_n/O_n).$$

Here $\Gamma(M; PL_n/O_n)$ denotes the space of sections $s$ in the fiber bundle associated to $P \to M$ with fiber $PL_n/O_n$, such that $s|\partial M$ maps to the base point in each fiber. (The precise statement requires a detour via spaces of piecewise differentiable maps, which we suppress.) For concordance spaces [BL77, (2.4)] there is a similar homotopy fiber sequence

$$(1.4.3) \qquad C^{DIFF}(M) \to C^{PL}(M) \to \Gamma(M; C_n),$$

where $\Gamma(M; C_n)$ is the space of sections in a bundle over $M$ with fiber

$$C_n = \mathrm{hofib}(PL_n/O_n \to PL_{n+1}/O_{n+1}),$$

with prescribed behavior on $\partial M$. (Burghelea–Lashof use the notation $F_n$ for the $TOP/DIFF$ analog of this homotopy fiber.)

Let

$$F^{CAT}(M) = \mathrm{hofib}(C^{CAT}(M) \xrightarrow{\sigma} C^{CAT}(M \times J))$$

be the homotopy fiber of the suspension map for CAT concordances. By [BL77, Thm. A] the concordance suspension maps are compatible with a suspension map $\varphi \colon C_n \to \Omega C_{n+1}$, so there is a homotopy fiber sequence

(1.4.4)                 $$F^{DIFF}(M) \to F^{PL}(M) \to \Gamma(M; F_n),$$

where $\Gamma(M; F_n)$ is the space of sections in a bundle with fiber

$$F_n = \mathrm{hofib}(C_n \xrightarrow{\varphi} \Omega C_{n+1}),$$

still with prescribed behavior on $\partial M$.

The columns in the following diagram are homotopy fiber sequences:

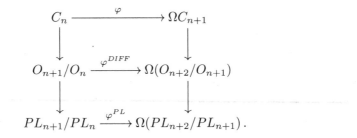

The lower vertical arrows are $(n+2)$-connected for $n \geq 5$, by the $PL/DIFF$ stability theorem [KS77, V.5.2]. Hence $C_n$, $\Omega C_{n+1}$ and the upper horizontal map $\varphi$ are all $(n+1)$-connected, and the homotopy fiber $F_n$ is at least $n$-connected.

Igusa's theorem implies that $F^{DIFF}(M)$ is $(k-1)$-connected, for $M$, $n$ and $k$ as in the statement of the corollary. In addition, $\sigma \colon C^{DIFF}(M) \to C^{DIFF}(M \times J)$ is 0-connected (for $n \geq 7$). Since $\Gamma(M; \Omega C_{n+1})$ is 1-connected, it follows from (1.4.3) that $\sigma \colon C^{PL}(M) \to C^{PL}(M \times J)$ is at least 0-connected.

Now consider the special case $M = D^n$. The spaces $PL(D^n)$, $C^{PL}(D^n)$ and $F^{PL}(D^n)$ are all contractible, by the Alexander trick. The tangent bundle of $D^n$ is trivial, so $\Gamma(D^n; F_n) = \Omega^n F_n$. Igusa's theorem and (1.4.4) then imply that $\Omega^n F_n$ is $k$-connected. It follows that $F_n$ is $(n+k)$-connected, because we saw from the $PL/DIFF$ stability theorem that $F_n$ is at least $n$-connected.

Returning to the case of a general smoothable $n$-manifold $M$, the section space $\Gamma(M; F_n)$ is $k$-connected by obstruction theory. Hence $F^{PL}(M)$ is $(k-1)$-connected by Igusa's theorem and (1.4.4). It follows that the PL concordance stabilization map $\sigma\colon C^{PL}(M) \to C^{PL}(M \times J)$ is $k$-connected, because we saw from Igusa's theorem and (1.4.3) that it is at least 0-connected. $\square$

In a relative way, Igusa's theorem improves on the cited $PL/DIFF$ stability theorem, by showing that the PL suspension map

$$PL_{n+1}/PL_n \xrightarrow{\varphi^{PL}} \Omega(PL_{n+2}/PL_{n+1})$$

is at least $(n+k+2)$-connected, when $n \geq \max\{2k+7, 3k+4\}$. For comparison, the DIFF suspension map

$$S^n \cong O_{n+1}/O_n \xrightarrow{\varphi^{DIFF}} \Omega(O_{n+2}/O_{n+1}) \cong \Omega S^{n+1}$$

is precisely $(2n - 1)$-connected, by Freudenthal's theorem.

*Remark 1.4.5.* There are some forward references in [Wa85, §3.1] concerning simple maps to (an earlier version of) the present work. For the reader's convenience, we make these explicit here. The claim that simple maps form a category, and satisfy a gluing lemma [Wa85, p. 401] is contained in our Proposition 2.1.3. The claim that the $H$-space $s\mathcal{C}^h(X)$ is grouplike [Wa85, p. 402] is our Corollary 3.2.4.

The proof of [Wa85, Lem. 3.1.4] contains three forward references. The "well known argument" was implicit in [GZ67], and is made explicit in our Lemma 3.2.14. The result that the last vertex map is simple is our Proposition 2.2.18. The fact that subdivision preserves simple maps is our Proposition 2.3.3. In our proof, the full strength of our Theorem 2.3.2 is used. Thus that result, on the quasi-naturality of the Fritsch–Puppe homeomorphism, is presently required for the identification $s\mathcal{C}^h(X) \simeq \Omega \mathrm{Wh}^{PL}(X)$, and thus for Theorems 0.1, 0.2 and 0.3.

On top of page 405 of [Wa85], use is made of a simplicial deformation retraction of $[n] \mapsto X^{\Delta^n \times \Delta^n}$ onto $[n] \mapsto X^{\Delta^n}$, where $X$ is a simplicial set. The relevant inclusion is induced by the projection $pr_1\colon \Delta^n \times \Delta^n \to \Delta^n$, and we take the retraction to be induced by the diagonal map $diag\colon \Delta^n \to \Delta^n \times \Delta^n$. Then $diag \circ pr_1$ is the nerve of the composite functor (= order-preserving function) $f\colon [n] \times [n] \to [n] \times [n]$ that takes $(i,j)$ to $(i,i)$, for $i,j \in [n]$. There is a chain of natural transformations

$$(i,i) \leq (i, \max(i,j)) \geq (i,j)$$

relating $f$ to the identity on $[n] \times [n]$, which is natural in $[n]$. Taking nerves, we get a chain of simplicial homotopies relating $diag \circ pr_1$ to the identity on $\Delta^n \times \Delta^n$, which is still natural in $[n]$. Forming mapping spaces into $X$, we obtain the required chain of simplicial homotopies.

There are two references on page 406 of [Wa85] to our Proposition 3.2.5, i.e., the fact that the functor $X \mapsto s\mathcal{C}^h(X)$ respects weak homotopy equivalences.

The concluding reformulation [Wa85, Prop. 3.3.2] of [Wa85, Thm. 3.3.1] is not correct as stated. In the definition of the simplicial category $\mathcal{R}_f(X)_\bullet$, the condition on the objects in simplicial degree $q$, that the composite map

$$Y \xrightarrow{r} X \times \Delta^q \xrightarrow{pr} \Delta^q$$

is locally fiber homotopy trivial, should be replaced with the stronger condition that the map is a Serre fibration. This leads to the following definition and corrected proposition.

**Definition 1.4.6.** To each simplicial set $X$ we associate a simplicial category $\widetilde{\mathcal{R}}_\bullet(X)$. In simplicial degree $q$, it is the full subcategory of $\mathcal{R}_f(X \times \Delta^q)$ generated by the objects $(Y, r, y)$ for which the composite map

$$Y \xrightarrow{r} X \times \Delta^q \xrightarrow{pr} \Delta^q$$

is a Serre fibration. Let $\widetilde{\mathcal{R}}_\bullet^h(X)$ be the full simplicial subcategory with objects such that $y \colon X \times \Delta^q \to Y$ is also a weak homotopy equivalence, and let $s$- and $h$-prefixes indicate the subcategories of simple maps and weak homotopy equivalences, respectively.

By [Wa85, Thm. 3.3.1], there is a homotopy cartesian square

(1.4.7)
$$
\begin{array}{ccc}
sS_\bullet \mathcal{R}_f^h(X^{\Delta^\bullet}) & \longrightarrow & sS_\bullet \mathcal{R}_f(X^{\Delta^\bullet}) \\
\downarrow & & \downarrow \\
hS_\bullet \mathcal{R}_f^h(X^{\Delta^\bullet}) & \longrightarrow & hS_\bullet \mathcal{R}_f(X^{\Delta^\bullet})
\end{array}
$$

where the entries have the following meaning. Let $X^K$ denote the mapping space $\mathrm{Map}(K, X)$, with $p$-simplices the maps $\Delta^p \times K \to X$. For each $q$, let $X^{\Delta^q}$ denote this mapping space (that is, take $K = \Delta^q$). Then the entries in the diagram need to be taken in the following slightly tricky sense: for each fixed $q$ evaluate the functor in question on $X^{\Delta^q}$, and then take the simplicial object that results by varying $q$.

The upper left hand term is one model for $\mathrm{Wh}^{PL}(X)$, the lower left hand term is contractible, and the loop spaces of the right hand terms are homotopy equivalent to $h(X; A(*))$ and $A(X)$, respectively. The homotopy fiber sequence (1.3.3) is part of the Puppe fiber sequence derived from this homotopy cartesian square.

**Proposition 1.4.8.** *There is a homotopy cartesian square*

$$
\begin{array}{ccc}
sS_\bullet \widetilde{\mathcal{R}}_\bullet^h(X) & \longrightarrow & sS_\bullet \widetilde{\mathcal{R}}_\bullet(X) \\
\downarrow & & \downarrow \\
hS_\bullet \widetilde{\mathcal{R}}_\bullet^h(X) & \longrightarrow & hS_\bullet \widetilde{\mathcal{R}}_\bullet(X)
\end{array}
$$

*and it is homotopy equivalent to the square (1.4.7) by a natural map.*

*Proof.* The natural map of homotopy cartesian squares is induced by the functor $\mathcal{R}_f(X^{\Delta^\bullet}) \to \widetilde{\mathcal{R}}_\bullet(X)$ given in simplicial degree $q$ by the composite

$$\mathcal{R}_f(X^{\Delta^q}) \to \mathcal{R}_f(X^{\Delta^q} \times \Delta^q) \to \mathcal{R}_f(X \times \Delta^q),$$

where the first map is given by product with $\Delta^q$, and the second map is functorially induced by

$$(ev, pr)\colon X^{\Delta^q} \times \Delta^q \to X \times \Delta^q$$

where $ev$ is the evaluation map. When applied to a retractive space $Y$ over $X^{\Delta^q}$ the result is a Serre fibration over $X \times \Delta^q$, by the fiber gluing lemma for Serre fibrations, Lemma 2.7.10. The proof then proceeds as in [Wa85, p. 418], up to the claim that $s\mathcal{C}^h(X) \to s\widetilde{\mathcal{C}}^h_\bullet(X)$ is a homotopy equivalence. This will be proved as Corollary 3.5.2.

Actually, the fiber gluing lemma for Serre fibrations was also used in the verification that the simplicial category $\widetilde{\mathcal{R}}_\bullet(X)$ in Definition 1.4.6 is a simplicial category with cofibrations; in particular, that for every $q$ the category $\widetilde{\mathcal{R}}_q(X)$ is a category with cofibrations, as is required for the use of the $S_\bullet$-construction. $\square$

In the "manifold approach" paper [Wa82, Prop. 5.1], a similar (approximately) homotopy cartesian square is constructed, where the entries are simplicial (sets or) categories of CAT manifolds. As discussed in [Wa82, pp. 178–180], there is a chain of homotopy equivalences relating the manifold square to the square of Proposition 1.4.8. This is how one can deduce [Wa82, Prop. 5.5], for CAT $=$ PL, asserting that the manifold functor that corresponds to

$$sS_\bullet\mathcal{R}_f(X^{\Delta^\bullet}) \simeq sS_\bullet\widetilde{\mathcal{R}}_\bullet(X)$$

is indeed a homological functor in $X$.

# Chapter Two

# On simple maps

## 2.1. SIMPLE MAPS OF SIMPLICIAL SETS

In this section we define simple maps of finite simplicial sets, and establish some of their formal properties.

Let $\Delta$ be the skeleton category of finite non-empty ordinals, with objects the linearly ordered sets $[n] = \{0 < 1 < \cdots < n\}$ for $n \geq 0$, and morphisms $\alpha \colon [n] \to [m]$ the order-preserving functions. A simplicial set $X$ is a contravariant functor from $\Delta$ to sets. The simplicial set $\Delta^q$ is the functor represented by the object $[q]$. By a simplex in $X$ we mean a pair $([n], x)$, where $n \geq 0$ and $x \in X_n$, but we shall usually denote it by $x$, leaving the simplicial degree $n$ implicit. We refer to [GZ67], [FP90, §4] or [GJ99] for the theory of simplicial sets.

**Definition 2.1.1.** Let $f \colon X \to Y$ be a map of finite simplicial sets. We say that $f$ is a **simple map** if its geometric realization $|f| \colon |X| \to |Y|$ has contractible point inverses, i.e., if for each point $p \in |Y|$ the preimage $|f|^{-1}(p)$ is a contractible topological space.

We sometimes denote a simple map by $X \xrightarrow{\simeq_s} Y$.

**Proposition 2.1.2.** *A map $f \colon X \to Y$ of finite simplicial sets is a simple map if and only if $|f| \colon |X| \to |Y|$ is a hereditary weak homotopy equivalence, i.e., if for each open subset $U \subset |Y|$ the restricted map $|f|^{-1}(U) \to U$ is a weak homotopy equivalence. In particular, a simple map is a weak homotopy equivalence.*

This will follow immediately from the more detailed Proposition 2.1.8, which in turn suggests the more general Definition 2.6.3 (when $X$ and $Y$ are arbitrary topological spaces). We shall principally use the following consequences.

**Proposition 2.1.3.** *Let $f \colon X \to Y$ and $g \colon Y \to Z$ be maps of finite simplicial sets.*

*(a) (Composition) If $f$ and $g$ are simple, then so is the composite map $gf \colon X \to Z$.*

*(b) (Right cancellation) If $f$ and $gf$ are simple, then $g$ is simple.*

*(c) (Pullback) Pullbacks of simple maps are simple.*

*(d) (Gluing lemma) Let*

$$X_1 \longleftarrow X_0 \longrightarrow X_2$$
$$\downarrow \simeq_s \qquad \downarrow \simeq_s \qquad \downarrow \simeq_s$$
$$Y_1 \longleftarrow Y_0 \longrightarrow Y_2$$

*be a commutative diagram of finite simplicial sets, such that the maps $X_0 \to X_1$ and $Y_0 \to Y_1$ are cofibrations and the maps $X_i \to Y_i$ for $i = 0, 1, 2$ are simple. Then the induced map of pushouts*

$$X_1 \cup_{X_0} X_2 \to Y_1 \cup_{Y_0} Y_2$$

*is simple.*

Simple maps do not have the left cancellation property, as the composite $\Delta^0 \to \Delta^1 \to \Delta^0$ illustrates. The cofibration hypotheses in the gluing lemma cannot be omitted, as the vertical map from $\Delta^1 \supset \partial\Delta^1 \subset \Delta^1$ to $\Delta^0 \leftarrow \partial\Delta^1 \to \Delta^0$ shows.

Weak homotopy equivalences have the 2-out-of-3 property, which combines the composition, right cancellation and left cancellation properties. They are often not preserved under pullback, but satisfy the gluing lemma [GJ99, II.8.8].

*Proof of Proposition 2.1.3.* (a) For each open subset $U \subset |Z|$ the restricted map $|gf|^{-1}(U) \to U$ factors as

$$|gf|^{-1}(U) \xrightarrow{\simeq} |g|^{-1}(U) \xrightarrow{\simeq} U$$

where the first map is a weak homotopy equivalence because $f$ is simple and $|g|^{-1}(U)$ is open, and the second map is a weak homotopy equivalence because $g$ is simple and $U$ is open.

(b) For each open subset $U \subset |Z|$ the preimage $|g|^{-1}(U)$ is open in $|Y|$, so the left hand map and the composite map in

$$|gf|^{-1}(U) \xrightarrow{\simeq} |g|^{-1}(U) \to U$$

are both weak homotopy equivalences. Hence so is the right hand map.

(c) Geometric realization commutes with pullbacks, so this follows from the definition of simple maps in terms of preimages of points.

(d) Let
$$T(X_0 \to X_2) = X_0 \times \Delta^1 \cup_{X_0} X_2$$

be the (ordinary) mapping cylinder of the map $X_0 \to X_2$. The front inclusion $X_0 \to T(X_0 \to X_2)$ is a cofibration, and its composite with the cylinder projection $pr \colon T(X_0 \to X_2) \to X_2$ equals the given map $X_0 \to X_2$. The cylinder projection is a simple map, because formation of mapping cylinders commutes with geometric realization, and each point inverse of $|T(X_0 \to X_2)| \to |X_2|$ is a cone, hence contractible. Using that $X_0 \to X_1$ is a cofibration, it follows easily that also the map of pushouts $X_1 \cup_{X_0} T(X_0 \to X_2) \to X_1 \cup_{X_0} X_2$ is simple. Arguing likewise with the $Y_i$, we obtain a commutative square

$$
\begin{array}{ccc}
X_1 \cup_{X_0} T(X_0 \to X_2) & \xrightarrow{\;\simeq_s\;} & X_1 \cup_{X_0} X_2 \\
{\scriptstyle \simeq_s}\big\downarrow & & \big\downarrow \\
Y_1 \cup_{Y_0} T(Y_0 \to Y_2) & \xrightarrow{\;\simeq_s\;} & Y_1 \cup_{Y_0} Y_2
\end{array}
$$

with simple horizontal maps. The left hand vertical map is simple, for each point inverse of its geometric realization is also a point inverse of one of the maps $|X_i| \to |Y_i|$, for $i = 0, 1, 2$. By the composition and right cancellation properties of simple maps, it follows that the right hand vertical map is simple.  $\square$

The geometric realization $|X|$ of a simplicial set is a CW complex, with one open $n$-cell $e^n \subset |X|$ for each non-degenerate $n$-simplex in $X$. Recall that by the Yoneda lemma, each $n$-simplex $x \in X_n$ corresponds to a representing map $\bar{x} \colon \Delta^n \to X$, and conversely.

**Lemma 2.1.4.** *Let $f \colon X \to Y$ be a map of finite simplicial sets, let $p \in |Y|$ and let $e^m \subset |Y|$ be the open cell that contains $p$, in the CW structure on $|Y|$. The preimage $|f|^{-1}(p)$ is a finite CW complex, with $(n - m)$-skeleton equal to the intersection of $|f|^{-1}(p)$ with the $n$-skeleton of $|X|$. More precisely, $|f|^{-1}(p)$ has one open $(n - m)$-cell $|f|^{-1}(p) \cap e^n$ for each open $n$-cell $e^n$ that maps to $e^m$ under $f$, in the CW structure on $|X|$.*

*Proof.* Let $y \in Y_m$ be the non-degenerate simplex corresponding to $e^m$, and let $\xi \in |\Delta^m|$ be the unique interior point that is mapped to $p$ under $|\bar{y}| \colon |\Delta^m| \to |Y|$. Write $X = \Delta^n \cup_{\partial \Delta^n} X'$, with $\bar{x} \colon \Delta^n \to X$ representing a non-degenerate simplex $x \in X_n$, and suppose inductively that the lemma holds for the restricted map $f' \colon X' \to Y$. If $f(x) = \rho^*(y)$ for some degeneracy operator $\rho$, i.e., an order-preserving surjection $\rho \colon [n] \to [m]$, then we have a commutative diagram:

$$
\begin{array}{ccccc}
\partial \Delta^n & \rightarrowtail & \Delta^n & \xrightarrow{\;\rho\;} & \Delta^m \\
\big\downarrow & & {\scriptstyle \bar{x}}\big\downarrow & & \big\downarrow{\scriptstyle \bar{y}} \\
X' & \rightarrowtail & X & \xrightarrow[\;f\;]{} & Y.
\end{array}
$$

The preimage $D = |\rho|^{-1}(\xi)$ of the interior point $\xi$ under the affine linear map $|\rho| \colon |\Delta^n| \to |\Delta^m|$ is a convex $(n - m)$-cell, with boundary equal to the intersection $S = |\partial \Delta^n| \cap D$. By induction, $S$ is mapped into the $(n - m - 1)$-skeleton of $|f'|^{-1}(p)$, and $|f|^{-1}(p)$ is obtained from $|f'|^{-1}(p)$ by attaching $(D, S) \cong (D^{n-m}, S^{n-m-1})$ along this map. Hence $|X|$ has the asserted CW structure. If $f(x)$ is not a degeneration of $y$, then $|f'|^{-1}(p) = |f|^{-1}(p)$ and there is nothing more to prove.  $\square$

We shall use the following two results of R. C. Lacher. To state them, we first review some point-set topology.

**Definition 2.1.5.** A topological space is called a **Euclidean neighborhood retract** (ENR) if it is homeomorphic to a retract of an open subset of some (always finite-dimensional) Euclidean space. Every finite CW complex is a compact ENR [Ha02, Cor. A.10]. A metrizable space $X$ is called an **absolute neighborhood retract** (ANR) if, whenever $X$ is embedded as a closed subspace of a metric space $Z$, then $X$ is a retract of some neighborhood of $X$ in $Z$. Euclidean spaces are ANRs, by Tietze's extension theorem, and it is a formality that the class of ANRs is closed under passage to open subsets and retracts [FP90, Prop. A.6.4]. Therefore every ENR is an ANR.

The **Čech homotopy type** (or **shape**) of a compact subspace $X$ of the Hilbert space of square-summable sequences was defined by K. Borsuk [Bo68, §8]. By Urysohn's embedding theorem [Ur24], the class of these spaces equals the class of compact metrizable spaces, and includes all compact ENRs. Two compact ANRs embedded in Hilbert space have the same Čech homotopy type if and only if they have the same homotopy type [Bo68, (8.6)]. In particular, a compact ENR has the Čech homotopy type of a point if and only if it is contractible.

For separable metric spaces the Menger–Urysohn inductive dimension and the Čech–Lebesgue covering dimension are equal. See Definitions III.1 and V.1 and Theorem V.8 of Hurewicz–Wallman [HW41]. Every finite-dimensional separable metric space can be embedded in a Euclidean space, and conversely, every subspace of a Euclidean space is a finite-dimensional separable metric space. See Theorems III.1, IV.1 and V.3 of [HW41]. In particular, every ENR is a finite-dimensional separable metrizable space.

M. Brown [Br60] defined a subspace $A$ of an $n$-manifold $M$ to be **cellular** if it is the intersection $A = \bigcap_{i=1}^{\infty} Q_i$ of a sequence of closed, topological $n$-cells in $M$, with $Q_{i+1}$ contained in the interior of $Q_i$ for each $i \geq 1$.

R. C. Lacher [La69b, p. 718] defined a space $A$ to be **cell-like** if it can be embedded as a cellular subspace of some manifold. Furthermore, a map $f \colon X \to Y$ of topological spaces is **cell-like** if $f^{-1}(p)$ is a cell-like space for each $p \in Y$. Any cellular or cell-like space is non-empty, and any cell-like map is surjective.

D. R. McMillan [Mc64] defined an embedding $\varphi \colon A \to X$ to have **property $UV^{\infty}$** if for each open subset $U$ of $X$ containing $\varphi(A)$ there exists an open subset $V$ of $X$ with $\varphi(A) \subset V \subset U$, such that the inclusion map $V \subset U$ is null-homotopic in $U$.

A map $f \colon X \to Y$ of topological spaces is a **proper homotopy equivalence** if $f$ is proper (preimages of compact sets are compact) and there are proper maps $g \colon Y \to X$, $H \colon X \times I \to X$ and $H' \colon Y \times I \to Y$ such that $H$ is a homotopy from $gf$ to $id_X$ and $H'$ is a homotopy from $fg$ to $id_Y$.

**Theorem 2.1.6 (Lacher).** *The following conditions are equivalent for finite-dimensional compact metrizable spaces $A$:*

*(a) $A$ is cell-like;*

*(b) $A$ has the Čech homotopy type of a point;*

*(c) $A$ is nonempty and admits an embedding with property $UV^{\infty}$ into an ENR.*

*Proof.* This is the theorem on page 599 of [La69a], in the formulation given in [La69b, Thm. 1.1]. Obviously, either one of conditions (a) and (b) implies that $A$ is nonempty. $\square$

**Theorem 2.1.7 (Lacher).** *Let $f: X \to Y$ be a proper map of ENRs. Then (a) $f$ is cell-like if and only if (b) $f$ is surjective and for each open subset $U \subset Y$ the restricted map $f^{-1}(U) \to U$ is a proper homotopy equivalence.*

*Proof.* See [La69b, Thm. 1.2]. $\square$

**Proposition 2.1.8.** *Let $f: X \to Y$ be a map of finite simplicial sets, with geometric realization $|f|: |X| \to |Y|$. The following conditions (a)–(f) are equivalent:*

*(a) $f$ is a simple map;*

*(b) For each point $p \in |Y|$, the preimage $|f|^{-1}(p)$ has the Čech homotopy type of a point;*

*(c) $|f|$ is a cell-like mapping;*

*(d) $|f|$ is a hereditary proper homotopy equivalence, i.e., for each open subset $U \subset |Y|$ the restricted map $|f|^{-1}(U) \to U$ is a proper homotopy equivalence;*

*(e) $|f|$ is a hereditary homotopy equivalence, i.e., for each open subset $U \subset |Y|$ the restricted map $|f|^{-1}(U) \to U$ is a homotopy equivalence;*

*(f) $|f|$ is a hereditary weak homotopy equivalence.*

*Proof.* By Lemma 2.1.4, each preimage $|f|^{-1}(p)$ is a finite CW complex, hence a compact ENR. So by Borsuk's result cited above, $|f|^{-1}(p)$ is contractible if and only if it has the Čech homotopy type of a point. Thus $(a) \iff (b)$. By the dimension theory reviewed above, each $|f|^{-1}(p)$ is a finite-dimensional compact metrizable space. Thus $(b) \iff (c)$ by the equivalence of Theorem 2.1.6(a) and (b). The implication $(c) \implies (d)$ follows from Theorem 2.1.7 applied to the map $|f|: |X| \to |Y|$, which is proper (and closed) because $|X|$ is compact and $|Y|$ is Hausdorff. The implications $(d) \implies (e)$ and $(e) \implies (f)$ are obvious, because every proper homotopy equivalence is a homotopy equivalence, and every homotopy equivalence is a weak homotopy equivalence.

It remains to prove that $(f) \implies (c)$. Thus suppose that $f: X \to Y$ is a map of finite simplicial sets, such that $|f|: |X| \to |Y|$ is a hereditary weak homotopy equivalence. We shall demonstrate below that $|f|$ is surjective, and that the inclusion $A = |f|^{-1}(p) \subset |X|$ has property $UV^\infty$, for each $p \in |Y|$. This will complete the proof, because $|X|$ is an ENR, so by the equivalence of Theorem 2.1.6(a) and (c) each point inverse $|f|^{-1}(p)$ is cell-like, and this verifies (c).

The image $L = |f|(|X|) \subset |Y|$ is closed, because $|f|$ is closed, so $U = |Y| \setminus L$ is open. Its preimage $|f|^{-1}(U)$ is empty, so the restricted map $|f|^{-1}(U) \to U$ can only be a weak homotopy equivalence when $U$ is empty, i.e., when $|f|$ is surjective.

It remains to verify the property $UV^\infty$. Let $A = |f|^{-1}(p) \subset U \subset |X|$ with $U$ open. The complement $K = |X| \setminus U$ is closed, so its image $|f|(K) \subset |Y|$ is closed and does not contain $p$. Each point in a CW-complex has arbitrarily small contractible open neighborhoods [Ha02, Prop. A.4], so $p \in |Y|$ has a contractible open neighborhood $N$ that does not meet $|f|(K)$. By assumption, the restricted map $|f|^{-1}(N) \to N$ is a weak homotopy equivalence, so by defining $V = |f|^{-1}(N)$ we have obtained an open, weakly contractible neighborhood of $A$ that is contained in $U$, namely $V$.

Now $|X|$ is an ENR, and thus an ANR, so its open subset $V$ is also an ANR. Thus $V$ has the homotopy type of a CW-complex, by Milnor's theorem [Mi59, Thm. 1(a) and (d)]. See alternatively [FP90, Thm. 5.2.1]. Hence the weakly contractible space $V$ is in fact contractible, so the inclusion $V \subset U$ is null-homotopic.  $\square$

*Remark 2.1.9.* There is a technical variant of Definition 2.1.1 and Proposition 2.1.2 that will also be needed, principally in Propositions 2.7.5 and 2.7.6. Let $Z$ be a fixed simplicial set and $V \subset |Z|$ a fixed open subset. A map $f \colon X \to Y$ of finite simplicial sets over $Z$, i.e., a map commuting with given structure maps $X \to Z$ and $Y \to Z$, will be called **simple over** $V$ if for every open subset $U \subset |Y|$ that is contained in the preimage of $V$ the restricted map $|f|^{-1}(U) \to U$ is a weak homotopy equivalence. Equivalently, thanks to Lacher, $f$ is simple over $V$ if and only if each point $p \in |Y|$ contained in the preimage of $V$ has contractible preimage under $|f|$. Propositions 2.1.3 and 2.1.8 both carry over to this situation, in the sense that all simplicial sets and maps can be taken over $Z$, and the term "simple" may be replaced throughout by "simple over $V$." To see this one modifies the above arguments in a straightforward way.

## 2.2. NORMAL SUBDIVISION OF SIMPLICIAL SETS

In this section we define the Barratt nerve $B(X)$ and the Kan normal subdivision $Sd(X)$ of a simplicial set $X$. We also define the last vertex map $d_X \colon Sd(X) \to X$, and show that it is a simple map for each finite simplicial set $X$.

We call each injective morphism $\mu$ in $\Delta$ a **face operator**, and each surjective morphism $\rho$ in $\Delta$ a **degeneracy operator**. We write $\iota_n \colon [n] \to [n]$ for the identity morphism of $[n]$, often thought of as an $n$-simplex of $\Delta^n$. A **proper** face or degeneracy operator is one that is not equal to the identity. The $i$-th face operator $\delta^i \colon [n-1] \to [n]$ maps no element to $i$, while the $j$-th degeneracy operator $\sigma^j \colon [n+1] \to [n]$ maps two elements to $j$. For any simplicial object $X$ and morphism $\alpha \colon [m] \to [n]$ in $\Delta$, we write $\alpha^* \colon X_n \to X_m$ for the induced morphism.

**Definition 2.2.1.** The **nerve** $N\mathcal{C}$ of a small category $\mathcal{C}$ is the simplicial set with $q$-simplices the set $N_q\mathcal{C}$ of functors $c \colon [q] \to \mathcal{C}$, or equivalently, the set of diagrams

$$c_0 \to c_1 \to \cdots \to c_q$$

in $\mathcal{C}$. The simplicial structure is given by right composition, so $\alpha^*(c) = c \circ \alpha$ for each morphism $\alpha$ in $\Delta$. More explicitly, the $i$-th face operator deletes the object $c_i$, and the $j$-th degeneracy operator inserts the identity morphism of $c_j$. A functor $F\colon \mathcal{C} \to \mathcal{D}$ induces a map $NF\colon N\mathcal{C} \to N\mathcal{D}$ of simplicial sets, taking the $q$-simplex $c$ above to $F \circ c$, i.e., the resulting diagram $F(c_0) \to F(c_1) \to \cdots \to F(c_q)$ in $\mathcal{D}$. A natural transformation $t\colon F \to G$ of functors $F, G\colon \mathcal{C} \to \mathcal{D}$ can be viewed as a functor $T\colon \mathcal{C} \times [1] \to \mathcal{D}$, and induces a simplicial homotopy $NT\colon N\mathcal{C} \times \Delta^1 \cong N(\mathcal{C} \times [1]) \to N\mathcal{D}$ from $NF$ to $NG$.

The nerve $N\mathcal{C}$ of a partially ordered set $C$ is the nerve of $C$ viewed as a small category, i.e., with one object for each element of $C$, a single morphism $a \to b$ if $a \leq b$ in the partial ordering, and no morphisms from $a$ to $b$ otherwise. Any order-preserving function $\varphi\colon C \to D$ then induces a map $N\varphi\colon N\mathcal{C} \to N\mathcal{D}$.

The trivial observation that each simplex of a simplicial set is the degeneration of some non-degenerate simplex can be sharpened to the following uniqueness statement, known as the **Eilenberg–Zilber lemma**.

**Lemma 2.2.2.** *Each simplex $x$ of a simplicial set $X$ has a unique decomposition in the form*

$$x = \rho^*(x^{\#})\,,$$

*where $x^{\#}$ is a non-degenerate simplex in $X$ and $\rho$ is a degeneracy operator.*

*Proof.* See [EZ50, (8.3)] or [FP90, Thm. 4.2.3]. $\square$

We shall call $x^{\#}$ the **non-degenerate part** of $X$. If $x$ is already non-degenerate, then of course $x = x^{\#}$ and $\rho$ is the identity, but otherwise $\rho$ is a proper degeneracy operator.

The following construction $B(X)$ was called the nerve functor $NX$ by Barratt [Ba56, §2], and the star functor $X^*$ in [FP90, p. 219]. We will instead call it the Barratt nerve of $X$.

**Definition 2.2.3.** For any simplicial set $X$, let $X^{\#} = \{x^{\#} \mid x \in X\}$ be the set of its non-degenerate simplices. Give $X^{\#}$ the partial ordering where $x \leq y$ if $x$ is a face of $y$, i.e., if $x = \mu^*(y)$ for some face operator $\mu$. Let the **Barratt nerve** $B(X) = N(X^{\#}, \leq)$ be the nerve of this partially ordered set. A $q$-simplex of $B(X)$ is a chain $(x_0 \leq \cdots \leq x_q)$ of non-degenerate simplices of $X$, where $x_i$ is a face of $x_{i+1}$ for each $0 \leq i < q$.

A map $f\colon X \to Y$ of simplicial sets induces a function $f^{\#}\colon X^{\#} \to Y^{\#}$, given by $f^{\#}(x) = f(x)^{\#}$. This function is order-preserving, because if $x$ is a face of $y$, then the non-degenerate part of $f(x)$ is a face of the non-degenerate part of $f(y)$. Hence there is an induced map of nerves $B(f)\colon B(X) \to B(Y)$, which makes the Barratt nerve a covariant functor.

The Barratt nerve functor has bad homotopy properties in general. For example, when $X = \Delta^n/\partial\Delta^n$ for $n \geq 1$, the non-degenerate 0-simplex is a face of the non-degenerate $n$-simplex, and $B(X) \cong \Delta^1$ is contractible.

**Example 2.2.4.** For $X = \Delta^n$, $X^{\#}$ is the set of all faces of $\Delta^n$. It equals the set of face operators $\mu\colon [m] \to [n]$ with target $[n]$, and is isomorphic to the set of non-empty subsets $\mu([m]) \subset [n]$, ordered by inclusion. When viewed as an ordered simplicial complex [FP90, p. 152], the Barratt nerve $B(\Delta^n)$ can be identified with the barycentric subdivision of $\Delta^n$ [FP90, p. 111]. More precisely, there is a canonical homeomorphism

$$h_n : |B(\Delta^n)| \xrightarrow{\cong} |\Delta^n|$$

that takes the 0-cell $(\mu)$ in $|B(\Delta^n)|$ corresponding to a face operator $\mu : [m] \to [n]$ to the **barycenter** $\langle \mu \rangle$ of the face $|\mu|(|\Delta^m|)$ of $|\Delta^n|$. The barycenter of $|\Delta^m|$ is the point $\beta = \langle \iota_m \rangle$ with barycentric coordinates $(t_0, \ldots, t_m)$ all equal to $1/(m+1)$, and $\langle \mu \rangle$ is its image under $|\mu| : |\Delta^m| \to |\Delta^n|$. Its barycentric coordinates are thus $(t_0, \ldots, t_n)$, where now $t_i = 1/(m+1)$ if $i$ is in the image of $\mu$, and $t_i = 0$ otherwise.

Furthermore, the homeomorphism $h_n$ maps each simplex in $|B(\Delta^n)|$ affine linearly to $|\Delta^n|$, in the sense that for each $q$-simplex $x = (\mu_0 \leq \cdots \leq \mu_q)$ in $B(\Delta^n)$ the maps

$$|\Delta^q| \xrightarrow{|\bar{x}|} |B(\Delta^n)| \xrightarrow{h_n} |\Delta^n|$$

take $(t_0, \ldots, t_q)$ in $|\Delta^q|$ to $u = \sum_{k=0}^q t_k(\mu_k)$ in $|B(\Delta^n)|$, and then to

$$h_n(u) = \sum_{k=0}^q t_k \langle \mu_k \rangle$$

in $|\Delta^n|$. These homeomorphisms are natural for face operators $\mu : [m] \to [n]$, in the sense that $|\mu| \circ h_m = h_n \circ |B(\mu)|$. It follows that if $X$ is the simplicial set associated to an ordered simplicial complex, then $B(X)$ is the simplicial set associated to its barycentric subdivision, and there is a canonical homeomorphism $h_X : |B(X)| \cong |X|$. See also Theorem 2.3.1.

*Remark 2.2.5.* The homeomorphisms $h_n$ of Example 2.2.4 are **not** natural for most degeneracy operators $\rho$. For instance, the square

$$
\begin{array}{ccc}
|B(\Delta^2)| & \xrightarrow{h_2} & |\Delta^2| \\
{\scriptstyle |B(\rho)|} \downarrow & & \downarrow {\scriptstyle |\rho|} \\
|B(\Delta^1)| & \xrightarrow{h_1} & |\Delta^1|
\end{array}
$$

does not commute for either of the two degeneracy maps $\rho = \sigma^0, \sigma^1 : \Delta^2 \to \Delta^1$. In fact, there does not exist any natural homeomorphism $h_X : |B(X)| \cong |X|$ for $X$ in a category of simplicial sets that contains these two degeneracy maps. The two 0-cells $(\delta^1)$ and $(\iota_2)$ of $|B(\Delta^2)|$ will have different images under the embedding

$$(|\sigma^0|, |\sigma^1|) \circ h_{\Delta^2} : |B(\Delta^2)| \to |\Delta^1| \times |\Delta^1|,$$

but their images under $(h_{\Delta^1} \circ |B(\sigma^0)|, h_{\Delta^1} \circ |B(\sigma^1)|)$ will be the same. See [FP67, p. 508] and [FP90, pp. 124–125].

We now turn to the left Kan extension of the Barratt nerve, namely the normal subdivision functor, which has much better homotopy properties. The following definition is from [GZ67, II.1.1].

**Definition 2.2.6.** To each simplicial set $X$ we associate the **simplex category** $\text{simp}(X)$, whose objects are the pairs $([n], x)$ with $n \geq 0$ and $x \in X_n$, and whose morphisms from $([m], y)$ to $([n], x)$ are the morphisms $\alpha \colon [m] \to [n]$ in $\Delta$ such that $\alpha^*(x) = y$. In other words, $\text{simp}(X) = \Upsilon/X$ is the left fiber at $X$ of the Yoneda embedding $\Upsilon$ of $\Delta$ into simplicial sets. A map $f \colon X \to Y$ induces a functor $\text{simp}(X) \to \text{simp}(Y)$ that takes $([n], x)$ to $([n], f(x))$, so $\text{simp}(-)$ is a functor from simplicial sets to small categories.

The rule $([n], x) \mapsto \Delta^n$ defines a functor from $\text{simp}(X)$ to simplicial sets, and the representing maps $\bar{x} \colon \Delta^n \to X$ combine to a natural isomorphism

$$\operatornamewithlimits{colim}_{\text{simp}(X)} \left( ([n], x) \mapsto \Delta^n \right) \overset{\cong}{\longrightarrow} X \, .$$

The left hand side can also be written as the identification space

$$\coprod_{n \geq 0} (X_n \times \Delta^n)/\sim \, ,$$

where $\sim$ is generated by the relation $(\alpha^*(x), \varphi) \sim (x, \alpha \varphi)$ for $\alpha \colon [m] \to [n]$, $x \in X_n$ and $\varphi \in \Delta_q^m$, $q \geq 0$. The following definition is due to D. Kan [Ka57, §7], see also [FP90, §§4.2 and 4.6].

**Definition 2.2.7.** Let $X$ be a simplicial set. The rules $([n], x) \mapsto B(\Delta^n)$ and $\alpha \mapsto B(\alpha)$ define a functor from $\text{simp}(X)$ to simplicial sets. The **normal subdivision** of $X$ is equal to the colimit

$$Sd(X) = \operatornamewithlimits{colim}_{\text{simp}(X)} \left( ([n], x) \mapsto B(\Delta^n) \right)$$

of this functor. In other words, $Sd$ is the left Kan extension [Ma71, X.3(10)] of the functor $[n] \mapsto B(\Delta^n)$ from $\Delta$ to simplicial sets, along the Yoneda embedding $\Upsilon$ of $\Delta$ into simplicial sets. We can also write

$$(2.2.8) \qquad Sd(X) = \coprod_{n \geq 0} (X_n \times B(\Delta^n))/\sim \, ,$$

where $\sim$ is generated by the relation $(\alpha^*(x), \varphi) \sim (x, B(\alpha)(\varphi))$ for $\alpha \colon [m] \to [n]$, $x \in X_n$ and $\varphi \in B(\Delta^m)_q$, $q \geq 0$.

For each map $f \colon X \to Y$ the functor $([n], x) \mapsto B(\Delta^n)$ extends over the functor $\text{simp}(X) \to \text{simp}(Y)$, so there is an induced map of colimits

$$Sd(f) \colon Sd(X) \to Sd(Y)$$

that makes $Sd$ a functor from simplicial sets to simplicial sets. The representing maps $\bar{x} \colon \Delta^n \to X$ induce maps $B(\bar{x}) \colon B(\Delta^n) \to B(X)$ that combine to define a natural map

$$b_X \colon Sd(X) \to B(X) \, ,$$

also known as the canonical map from a left Kan extension.

Recall from Definition 1.2.2 that a simplicial set $X$ is non-singular if for each non-degenerate simplex $x \in X^\#$ the representing map $\bar{x} \colon \Delta^n \to X$ is a cofibration of simplicial sets, or equivalently, if all its vertices are distinct. Any ordered simplicial complex is non-singular, when viewed as a simplicial set. The nerve $NC$ of any partially ordered set (but not of every category) is such an ordered simplicial complex, and thus a non-singular simplicial set. In particular, the Barratt nerve $B(X)$ of any simplicial set is a non-singular simplicial set.

**Lemma 2.2.9.** *The normal subdivision functor $Sd$ preserves cofibrations and all colimits of simplicial sets. If $X$ is finite, or non-singular, then so is $Sd(X)$.*

*Proof.* The preservation of colimits is formal, say from the formula (2.2.8). Suppose $X = \Delta^n \cup_{\partial \Delta^n} X'$. Then $Sd(X)$ is obtained from $Sd(X')$ by attaching $Sd(\Delta^n)$ along the simplicial subset $Sd(\partial \Delta^n)$, so $Sd(X') \to Sd(X)$ is a cofibration. Preservation of general cofibrations follows by a passage to colimits. See also [FP90, Cor. 4.2.9 and 4.2.11].

The Barratt nerve $B(\Delta^n) = Sd(\Delta^n)$ is finite and non-singular, and we have just seen that if $\partial \Delta^n \to X'$ is a cofibration, then so is $Sd(\partial \Delta^n) \to Sd(X')$. Hence if $X$ is finite, or non-singular, then so is $Sd(X)$, by induction over the non-degenerate simplices of $X$ and a passage to colimits. $\square$

**Lemma 2.2.10.** *The canonical map $b_X \colon Sd(X) \to B(X)$ is surjective, for each simplicial set $X$.*

*Proof.* Any $q$-simplex of $B(X)$ can be written as a chain $\mu_0^*(x) \leq \cdots \leq \mu_q^*(x) = x$ of non-degenerate simplices of $X$, with $x \in X_n$ and $\mu_0 \leq \cdots \leq \mu_q = \iota_n$. Then $(x, \mu_0 \leq \cdots \leq \mu_q)$ is a $q$-simplex of $Sd(X)$ that maps to the given $q$-simplex of $B(X)$. $\square$

**Lemma 2.2.11.** *If $X$ is non-singular then the canonical map $b_X \colon Sd(X) \to B(X)$ is an isomorphism, and conversely.*

*Proof.* In the special case $X = \Delta^n$, the simplex category $\mathrm{simp}(\Delta^n)$ has the terminal object $([n], \iota_n)$, so the colimit defining $Sd(\Delta^n)$ is canonically isomorphic to the value of the functor at that object, and $b_{\Delta^n} \colon Sd(\Delta^n) \to B(\Delta^n)$ equals the canonical isomorphism.

More generally, for a finite simplicial set $X$ write $X \cong \Delta^n \cup_{\partial \Delta^n} X'$, where $\bar{x} \colon \Delta^n \to X$ is the representing map of a non-degenerate simplex $x \in X_n$ of maximal dimension. Consider the commutative square

$$
\begin{array}{ccc}
Sd(\Delta^n) \cup_{Sd(\partial \Delta^n)} Sd(X') & \xrightarrow{\ \cong\ } & Sd(X) \\
\cong \downarrow & & \downarrow b_X \\
B(\Delta^n) \cup_{B(\partial \Delta^n)} B(X') & \longrightarrow & B(X).
\end{array}
$$

The left hand vertical map is an isomorphism, by induction on the dimension of $X$ and the number of non-degenerate $n$-simplices in $X$, and the special case $X = \Delta^n$. The upper horizontal map is an isomorphism, by Lemma 2.2.9 for the case of pushouts.

By the assumption that $X$ is non-singular, $\bar{x}$ restricts to a cofibration $\partial\Delta^n \to X'$, which induces an injective function $(\partial\Delta^n)^\# \to (X')^\#$. Viewing it as an inclusion, the partially ordered set $X^\#$ is isomorphic to the union of $(\Delta^n)^\#$ and $(X')^\#$ along their intersection $(\partial\Delta^n)^\#$. It follows that the nerve $B(X) = NX^\#$ is the union of the nerves $B(\Delta^n)$ and $B(X')$ along $B(\partial\Delta^n)$, i.e., that the lower horizontal map in the commutative square is an isomorphism. Hence the right hand vertical map $Sd(X) \to B(X)$ is an isomorphism in this case.

The case of infinite non-singular $X$ follows by passage to the filtered colimit over the finite simplicial subsets of $X$, ordered by inclusion, because both $Sd$ and $B$ respect such colimits.

For the converse implication (which we will not make any use of), suppose that $Sd(X) \to B(X)$ is an isomorphism, and consider a non-degenerate $n$-simplex $x$ of $X$. Let $Y \subset X$ be the simplicial subset generated by $x$, and write $Y = \Delta^n \cup_{\partial\Delta^n} Z$. In the commutative diagram

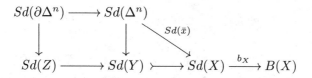

the square is a pushout and $Sd(Y) \to Sd(X)$ is a cofibration, by Lemma 2.2.9. As $\mu \colon [m] \to [n]$ ranges through the faces of $\Delta^n$, the pairwise distinct 1-simplices $(\mu \leq \iota_n)$ of $Sd(\Delta^n)$, which are not in $Sd(\partial\Delta^n)$, remain pairwise distinct in $Sd(Y)$ and in $Sd(X)$. Their images in $Sd(X)$ are the 1-simplices $(x, \mu \leq \iota_n)$. By the assumption on $b_X$ they remain pairwise distinct in $B(X)$, so the 1-simplices $(\mu^*(x)^\# \leq x)$ are all different. This means that the faces $\mu^*(x)$ of $x \in X_n$ are all non-degenerate and distinct, i.e., the representing map $\bar{x} \colon \Delta^n \to X$ is a cofibration. Thus $X$ is non-singular.   $\square$

*Remark 2.2.12.* For each simplicial set $X$ there is a universal/initial surjective map $X \to DX$ to a non-singular simplicial set $DX$, which we call the **desingularization** of $X$. This is because simplicial subsets, and arbitrary products, of non-singular simplicial sets are always non-singular, so $DX$ can be defined as the image of the canonical map

$$X \to \prod_{f \colon X \to Y} Y$$

that takes $x \in X_n$ to $(f(x))_f$, where $f$ ranges over all quotient maps from $X$ onto non-singular simplicial sets $Y$. Any surjective map $f \colon X \to Y$ to a non-singular $Y$ then factors uniquely over $X \to DX$.

Here are some open problems about desingularization: Can $DX$ be given a more explicit description? Is $D(Sd(X)) \cong B(X)$ for regular or op-regular $X$ (see Definition 2.5.3)? Is $D(T(f)) \cong M(f)$ when $f \colon X \to Y$ has simple cylinder reduction (see Definition 2.4.9)?

**Definition 2.2.13.** We define the cone on $\Delta^n$ to be $\mathrm{cone}(\Delta^n) = N([n] \cup \{v\})$, where $[n] \cup \{v\}$ is ordered by adjoining a new, greatest, element $v$ to $[n]$. The rule

$$(2.2.14) \qquad\qquad ([n], x) \mapsto \mathrm{cone}(\Delta^n)$$

defines a functor from $\mathrm{simp}(X)$ to simplicial sets. Some care is needed to define the cone on a simplicial set $X$ as a colimit, in the style of Definition 2.2.7, because we want the cone on the empty space $X = \emptyset$ to be a single point. Let $\mathrm{simp}_\eta(X)$ be the **augmented simplex category** obtained by adjoining an initial object $(-1, \eta)$ to $\mathrm{simp}(X)$, thought of as a unique $(-1)$-simplex in $X$. The rule $(2.2.14)$ extends to a functor from $\mathrm{simp}_\eta(X)$ to simplicial sets, taking the new object $(-1, \eta)$ to the vertex point $\mathrm{cone}(\Delta^{-1}) = N(\{v\})$. The **cone** on $X$ is then defined as

$$\mathrm{cone}(X) = \operatorname*{colim}_{\mathrm{simp}_\eta(X)} \left( ([n], x) \mapsto \mathrm{cone}(\Delta^n) \right).$$

The inclusion $[n] \subset [n] \cup \{v\}$ of partially ordered sets induces the natural **base inclusions** $\Delta^n \to \mathrm{cone}(\Delta^n)$ and $i \colon X \to \mathrm{cone}(X)$ of simplicial sets. In the source of $i$ it does not matter whether we form the colimit of $([n], x) \mapsto \Delta^n$ over $\mathrm{simp}(X)$ or $\mathrm{simp}_\eta(X)$, because the value $\Delta^{-1}$ of this functor on $(-1, \eta)$ is empty.

*Remark 2.2.15.* The $q$-simplices of $\mathrm{cone}(X)$ can be explicitly described as the pairs $(\mu \colon [p] \subset [q], x \in X_p)$ for $0 \le p \le q$, where $\mu = \delta^q \dots \delta^{p+1}$ is a **front face** of $[q]$, together with the vertex $v$. For more on $(-1)$-simplices, see Definition 2.4.1. We shall see in Lemma 2.4.11(a) that $\mathrm{cone}(X)$ is naturally isomorphic to the reduced mapping cylinder $M(f)$ of the unique map $f \colon X \to *$.

**Lemma 2.2.16.** *The cone functor preserves cofibrations. If $X$ is finite, or non-singular, then so is* $\mathrm{cone}(X)$.

*Proof.* Suppose $X = \Delta^n \cup_{\partial \Delta^n} X'$. Then $\mathrm{cone}(X)$ is obtained from $\mathrm{cone}(X')$ by attaching $\mathrm{cone}(\Delta^n) \cong \Delta^{n+1}$ along the horn $\mathrm{cone}(\partial \Delta^n) \cong \Lambda^{n+1}_{n+1}$, see Definition 3.2.1. The three claims follow from this observation. $\square$

The geometric realizations $|Sd(X)|$ and $|X|$ are homeomorphic, even if no natural homeomorphism exists. In the case of singular ($=$ not non-singular) $X$ this is a difficult fact due to Fritsch and Puppe [FP67], which we will extend in Section 2.3. However, there exists another relation between $Sd(X)$ and $X$, which is natural and also much easier to set up, namely the last vertex map $d_X \colon Sd(X) \to X$. We recall its definition from [Ka57, §7].

**Definition 2.2.17.** Let $d_n\colon (\Delta^n)^{\#} \to [n]$ be the function

$$(\mu\colon [m] \to [n]) \longmapsto \mu(m)$$

that takes each non-degenerate simplex in $\Delta^n$ to its last vertex. It is order-preserving, because the last vertex of any face of $\mu$ is less than or equal to the last vertex of $\mu$. It is natural in $[n]$, because for each morphism $\alpha\colon [n] \to [p]$ in $\Delta$ the last vertex of $\alpha\mu$ equals the last vertex of its non-degenerate part $(\alpha\mu)^{\#}$.

Passing to nerves, we obtain a last vertex map of simplicial sets

$$d_n\colon B(\Delta^n) \to \Delta^n\,,$$

which is natural in $[n]$ in the sense that for each $\alpha\colon [n] \to [p]$ the square

$$
\begin{array}{ccc}
B(\Delta^n) & \xrightarrow{\;d_n\;} & \Delta^n \\
{\scriptstyle B(\alpha)}\big\downarrow & & \big\downarrow{\scriptstyle \alpha} \\
B(\Delta^p) & \xrightarrow{\;d_p\;} & \Delta^p
\end{array}
$$

commutes. Hence $([n], x) \mapsto d_n\colon B(\Delta^n) \to \Delta^n$ defines a natural transformation of functors from $\mathrm{simp}(X)$ to simplicial sets. The induced map of colimits of these functors is by definition the **last vertex map**

$$d_X\colon Sd(X) \to X\,.$$

It is straightforward to check that $d_X$ is natural in the simplicial set $X$.

**Proposition 2.2.18.** *The last vertex map $d_X\colon Sd(X) \to X$ is simple, for each finite simplicial set $X$.*

*Proof.* We proceed by induction on the dimension $n$ of $X$ and the number of non-degenerate $n$-simplices of $X$. Choose an isomorphism $X \cong \Delta^n \cup_{\partial\Delta^n} X'$. In order to show that the map

$$d_X\colon Sd(X) \cong Sd(\Delta^n) \cup_{Sd(\partial\Delta^n)} Sd(X') \to \Delta^n \cup_{\partial\Delta^n} X' \cong X$$

is simple, it will suffice, by the gluing lemma for simple maps and Lemma 2.2.9, to show that the three maps $Sd(\Delta^n) \to \Delta^n$, $Sd(\partial\Delta^n) \to \partial\Delta^n$ and $Sd(X') \to X'$ are all simple. In the second and third cases this holds by the inductive hypothesis.

To handle the first case, we note that $\Delta^n$ has a greatest non-degenerate simplex (the identity $\iota_n$ on $[n]$), so $Sd(\Delta^n)$ may be identified with the cone on $Sd(\partial\Delta^n)$, and the map $d_{\Delta^n}$ may be factored as a composite

$$\mathrm{cone}(Sd(\partial\Delta^n)) \to \mathrm{cone}(\partial\Delta^n) \to \Delta^n\,.$$

The first map is the cone on the last vertex map for $\partial\Delta^n$, which is simple by the inductive hypothesis and the pullback and gluing properties of simple maps. The second map takes the cone point to the last vertex of $\Delta^n$, and its geometric realization has point inverses that are points or intervals, so it is also a simple map. $\square$

**Definition 2.2.19.** There is an involutive covariant functor $(-)^{op} \colon \Delta \to \Delta$ that, heuristically, takes each object $[n] = \{0 < 1 < \cdots < n\}$ to the same set with the reversed total ordering. Since we are working with $\Delta$ as a small skeleton for the category of all finite non-empty ordinals, $[n]^{op}$ is then canonically identified with $[n]$ again, but each morphism $\alpha \colon [m] \to [n]$ is mapped to the morphism $\alpha^{op}$ given by $\alpha^{op}(i) = n - \alpha(m - i)$ for $i \in [m]$.

To each simplicial set $X$, considered as a contravariant functor from $\Delta$ to sets, the **opposite** simplicial set $X^{op}$ is the composite functor $X \circ (-)^{op}$. The set of $n$-simplices of $X^{op}$ equals that of $X$, but the ordering of the vertices is reversed, in the sense that on $n$-simplices the $i$-th face map of $X^{op}$ equals the $(n - i)$-th face map of $X$, and similarly for the other simplicial structure maps. There is a natural homeomorphism of geometric realizations $|X| \cong |X^{op}|$, induced by the homeomorphism $|\Delta^n| \cong |\Delta^n|$ that takes $(t_0, \ldots, t_n)$ to $(t_n, \ldots, t_0)$ in barycentric coordinates.

**Definition 2.2.20.** The simplicial opposite of the normal subdivision $Sd(X)$ is the **op-normal subdivision** $Sd^{op}(X) = Sd(X)^{op}$. It can be defined as the colimit of the functor $([n], x) \mapsto B^{op}(\Delta^n) = B(\Delta^n)^{op}$, where now $B^{op}(\Delta^n)$ may be defined as the nerve of the partially ordered set of non-degenerate simplices in $\Delta^n$ with the opposite ordering from that of Definition 2.2.3, or inductively as the cone on $Sd^{op}(\partial \Delta^n)$, with the rule that the cone point is the **initial** vertex of any simplex containing it.

Let $d_n^{op} \colon (\Delta^n)^\# \to [n]$ be the function that takes each face $\mu \colon [m] \to [n]$ to its "first" vertex $\mu(0)$. It induces a map of nerves $d_n^{op} \colon B^{op}(\Delta^n) \to \Delta^n$, which is natural in $[n]$. For each simplicial set $X$ the rule $([n], x) \mapsto d_n^{op}$ defines a natural transformation of functors from $\mathrm{simp}(X)$ to simplicial sets, with colimit the **first vertex map**

$$d_X^{op} \colon Sd^{op}(X) \to X \, .$$

The normal subdivisions $Sd(X)$ and $Sd(X^{op})$ are in fact equal, as quotients of $\coprod_{n \geq 0} X_n \times B(\Delta^n) = \coprod_{n \geq 0} X_n^{op} \times B(\Delta^n)$. So the first vertex map for $X$ equals the opposite of the last vertex map for $X^{op}$. In particular, Proposition 2.2.18 implies that also the first vertex map $d_X^{op}$ is simple, for each finite simplicial set $X$.

## 2.3. Geometric realization and subdivision

The following Theorem 2.3.1, that the geometric realizations of $X$ and $Sd(X)$ are homeomorphic, was stated by Barratt in [Ba56, Thm. 3], but first properly proved in [FP67, Satz] by Fritsch and Puppe, because [Ba56, Lem. 3.1] is wrong. See also [FP90, Thm. 4.6.4]. We shall also obtain Theorem 2.3.2, asserting that the homeomorphisms $|Sd(X)| \cong |X|$ and $|Sd(Y)| \cong |Y|$ can be chosen to be compatible with respect to a previously given map $f \colon X \to Y$, when $Y$ is non-singular.

This kind of quasi-naturality statement lets us prove in Proposition 2.3.3 that $Sd$ preserves simple maps of finite simplicial sets. This result is needed for the proof of Lemma 3.1.4 and Theorem 3.1.7 of [Wa85], which in turn is needed for our homotopy fiber sequence (1.3.3), Theorems 0.2 and 0.3, and the DIFF case of Theorem 0.1. The quasi-naturality also implies that $Sd$ preserves Serre fibrations, which we make use of in the proof of Proposition 2.7.6. As discussed in Remark 2.2.5, there does not exist any natural homeomorphism $|Sd(X)| \cong |X|$.

In the special case when we only consider finite non-singular simplicial sets $X$ and $Y$, the proofs are much easier, and we shall give these first. This will suffice for the application of these results to Theorem 1.2.6. It will also suffice for the non-functorial first proof of Theorem 1.2.5, given in Section 3.1. Thereafter, we shall refine the proof of Fritsch and Puppe to also cover the case of general simplicial sets. This will be needed for the construction of the improvement functor $I$ in Section 2.5, and thus for the functorial second proof of Theorem 1.2.5, as well as for the cited results from [Wa85, §3.1].

**Theorem 2.3.1.** *Let $X$ be a simplicial set. There exists a homeomorphism*

$$h_X \colon |Sd(X)| \xrightarrow{\cong} |X|,$$

*which is homotopic to the geometric realization $|d_X|$ of the last vertex map.*

**Theorem 2.3.2.** *Let $f \colon X \to Y$ be a map of simplicial sets, with $Y$ non-singular. There exists a homeomorphism $h^f \colon |Sd(X)| \cong |X|$ such that the square*

$$
\begin{array}{ccc}
|Sd(X)| & \xrightarrow{\ h^f\ } & |X| \\
{\scriptstyle |Sd(f)|} \big\downarrow & & \big\downarrow {\scriptstyle |f|} \\
|Sd(Y)| & \xrightarrow{\ h_Y\ } & |Y|
\end{array}
$$

*commutes, with $h^f$ homotopic to $|d_X|$ and $h_Y \colon |Sd(Y)| \cong |Y|$ as in Theorem 2.3.1.*

We first make the following deduction.

**Proposition 2.3.3.** *Let $f \colon X \to Y$ be a map of finite simplicial sets. If $f$ is simple then so is its normal subdivision $Sd(f) \colon Sd(X) \to Sd(Y)$, and conversely.*

*Proof.* Suppose first that $f \colon X \to Y$ is a simple map of non-singular simplicial sets. Using only the non-singular case of Theorem 2.3.2, there are homeomorphisms $h^f \colon |Sd(X)| \cong |X|$ and $h_Y \colon |Sd(Y)| \cong |Y|$ such that $|f| \circ h^f = h_Y \circ |Sd(f)|$. Thus each preimage $|Sd(f)|^{-1}(p')$ for $p' \in |Sd(Y)|$ is homeomorphic to a preimage $|f|^{-1}(p)$ for $p = h_Y(p')$, and therefore contractible. So $Sd(f)$ is a simple map.

Suppose next that $f \colon X \to Y$ is a simple map, with $Y$ non-singular. By the general simplicial case of Theorem 2.3.2, the argument just given shows that $Sd(f)$ is a simple map.

Now suppose that $f: X \to Y$ is any simple map, and write $Y = \Delta^n \cup_{\partial \Delta^n} Y'$. Let $X \times_Y \Delta^n$ be the pullback of $\Delta^n \to Y$ along $f: X \to Y$, and similarly for $X \times_Y \partial \Delta^n$ and $X \times_Y Y'$. In the commutative diagram

$$\begin{array}{ccccc}
X \times_Y \Delta^n & \longleftarrow & X \times_Y \partial \Delta^n & \longrightarrow & X \times_Y Y' \\
\downarrow & & \downarrow & & \downarrow \\
\Delta^n & \longleftarrow & \partial \Delta^n & \longrightarrow & Y'
\end{array}$$

the vertical maps are simple by the pullback property, and the pushout in the upper row is isomorphic to $X \times_Y Y = X$. Apply $Sd$ to this diagram. The resulting vertical maps $Sd(X \times_Y \Delta^n) \to Sd(\Delta^n)$ and $Sd(X \times_Y \partial \Delta^n) \to Sd(\partial \Delta^n)$ are simple, by the case of a simple map to a non-singular simplicial set (namely $\Delta^n$ and $\partial \Delta^n$) just considered. The vertical map $Sd(X \times_Y Y') \to Sd(Y')$ is simple by an induction on the number of non-degenerate simplices in $Y$. Hence the pushout map $Sd(f): Sd(X) \to Sd(Y)$ is simple, by the gluing lemma for simple maps.

The proof of the converse is easier, and does not require Theorem 2.3.2. By naturality of the last vertex map we have $f \circ d_X = d_Y \circ Sd(f)$, where $d_X$ and $d_Y$ are simple by Proposition 2.2.18. Hence if $Sd(f)$ is simple, then $f$ is simple by the composition and right cancellation properties of simple maps.    □

We continue by reviewing the proof of Theorem 2.3.1 by Fritsch and Puppe, introducing some notation along the way. For the subsequent proof of Theorem 2.3.2, the main change will be that barycenters of simplices in $X$ have to be replaced by suitably defined pseudo-barycenters of simplices in $X$, with respect to $f$. In each case, we first discuss the easier proof in the case of finite non-singular $X$ (and $Y$).

*Proof of Theorem 2.3.1, the non-singular case.* Suppose that $X$ is a finite non-singular simplicial set. Then there is a homeomorphism $h_X: |Sd(X)| \to |X|$, affine linear on each simplex of $Sd(X)$, such that for each non-degenerate simplex $x$ of $X$, with representing map $\bar{x}: \Delta^n \to X$, the square

$$\begin{array}{ccc}
|Sd(\Delta^n)| & \xrightarrow{\ h_n\ } & |\Delta^n| \\
{\scriptstyle |Sd(\bar{x})|} \downarrow & & \downarrow {\scriptstyle |\bar{x}|} \\
|Sd(X)| & \xrightarrow{\ h_X\ } & |X|
\end{array}$$

commutes, where $h_n$ is as in Example 2.2.4. This property characterizes $h_X$ (if it exists), because the images of the maps $|Sd(\bar{x})|$ cover $|Sd(X)|$.

By induction, we can write $X = \Delta^n \cup_{\partial \Delta^n} X'$, where $\partial \Delta^n \to X'$ is a cofibration, and assume that $h_{\Delta^n} = h_n$, $h_{\partial \Delta^n}$ and $h_{X'}$ exist. To define $h_X$ as the induced map of pushouts in the diagram

$$|Sd(\Delta^n)| \longleftarrow |Sd(\partial\Delta^n)| \longrightarrow |Sd(X')|$$
$$h_n \downarrow \qquad\qquad h_{\partial\Delta^n} \downarrow \qquad\qquad h_{X'} \downarrow$$
$$|\Delta^n| \longleftarrow |\partial\Delta^n| \longrightarrow |X'|,$$

we just need to know that the left hand square commutes, because the right hand square commutes by the characterizing property. This follows from the fact that the homeomorphisms $h_n$ are natural for face operators, and all the non-degenerate simplices of $\partial\Delta^n$ are faces of $\Delta^n$. $\square$

*Proof of Theorem 2.3.1, the general case.* Now let $X$ be any simplicial set. For each simplex $x \in X_n$, Fritsch and Puppe [FP67] construct a map

$$(2.3.4) \qquad\qquad h_x \colon |B(\Delta^n)| \to |\Delta^n|$$

that depends on $([n], x)$ in $\mathrm{simp}(X)$, with the following naturality property.

**Proposition 2.3.5.** *For every morphism $\alpha \colon [n] \to [p]$ in $\Delta$, and pair of simplices $x \in X_n$, $y \in X_p$ with $\alpha^*(y) = x$, the square*

$$
\begin{array}{ccc}
|B(\Delta^n)| & \xrightarrow{\ h_x\ } & |\Delta^n| \\
{\scriptstyle |B(\alpha)|}\downarrow & & \downarrow{\scriptstyle |\alpha|} \\
|B(\Delta^p)| & \xrightarrow{\ h_y\ } & |\Delta^p|
\end{array}
$$

*commutes.*

The rule $([n], x) \mapsto h_x$ thus defines a natural transformation of functors from $\mathrm{simp}(X)$ to the category of topological spaces, whose induced map of colimits

$$h_X = \operatorname*{colim}_{\mathrm{simp}(X)} \big(([n], x) \mapsto h_x\big) \colon |Sd(X)| \to |X|$$

is the desired map $h_X$ for the theorem. Here we are using the fact that geometric realization commutes with the colimits expressing $X$ and $Sd(X)$, in Definitions 2.2.6 and 2.2.7, respectively. The map $h_X$ can also be characterized by the commutativity of the square

$$(2.3.6) \qquad
\begin{array}{ccc}
|B(\Delta^n)| & \xrightarrow{\ h_x\ } & |\Delta^n| \\
{\scriptstyle |Sd(\bar{x})|}\downarrow & & \downarrow{\scriptstyle |\bar{x}|} \\
|Sd(X)| & \xrightarrow{\ h_X\ } & |X|
\end{array}
$$

for each simplex $x \in X_n$, with representing map $\bar{x} \colon \Delta^n \to X$. In the upper left hand corner we have identified $Sd(\Delta^n) = B(\Delta^n)$. To show that $h_X$ is a homeomorphism, Fritsch and Puppe prove the following claim.

**Proposition 2.3.7.** *For each non-degenerate simplex $x \in X_n$, the map $h_x$ takes the interior of $|B(\Delta^n)|$ bijectively onto the interior of $|\Delta^n|$.*

It follows that $h_X$ is a continuous bijection. It is also closed, for $|Sd(X)|$ and $|X|$ have the topologies determined by all the maps $|Sd(\bar{x})|$ and $|\bar{x}|$, respectively, and Proposition 2.3.7 implies that $h_x$ is a closed surjection. Thus $h_X$ is a homeomorphism.

We now review the definition of the map $h_x$. Each point $u \in |B(\Delta^n)|$ is in the image of the affine $q$-simplex $|\Delta^q|$ associated to some $q$-simplex $(\mu_0 \leq \cdots \leq \mu_q)$ of $B(\Delta^n)$, and can therefore be written as a convex linear combination

$$(2.3.8) \qquad u = \sum_{k=0}^{q} t_k(\mu_k)$$

of the 0-cells $(\mu_k)$ in $|B(\Delta^n)|$. The numbers $(t_0, \ldots, t_q)$ with $\sum_{k=0}^{q} t_k = 1$ and $t_k \geq 0$ are the barycentric coordinates of $u$ in this $q$-simplex.

For each $0 \leq j \leq k \leq q$ the relation $\mu_j \leq \mu_k$ asserts that the image of $\mu_j$ is contained in the image of $\mu_k$, so there is a unique face operator $\mu_{kj}$ such that $\mu_j = \mu_k \mu_{kj}$. See diagram (2.3.10).

The placement of the simplex $x$ in the simplicial set $X$ enters as follows. For each $0 \leq k \leq q$ the face $\mu_k^*(x)$ can uniquely be written as $\rho_k^*(z_k)$ for some degeneracy operator $\rho_k$ and a non-degenerate simplex $z_k$ in $X$, by the Eilenberg–Zilber Lemma 2.2.2.

For each $0 \leq j \leq k \leq q$ we can uniquely factor the composite morphism $\rho_k \mu_{kj}$ in $\Delta$ as a degeneracy operator $\rho_{kj}$ followed by a face operator $\tilde{\mu}_{kj}$. We shall apply the following definition to the degeneracy operator $\rho_{kj}$.

**Definition 2.3.9.** For each degeneracy operator $\rho \colon [n] \to [m]$ we define its **maximal section** $\hat{\rho} \colon [m] \to [n]$ to be the face operator given by

$$\hat{\rho}(i) = \max \rho^{-1}(i) = \max\{j \in [n] \mid \rho(j) = i\}.$$

Then $\rho\hat{\rho} = \iota_m$ is the identity on $[m]$. For two composable degeneracy operators $\sigma$ and $\tau$ the maximal sections satisfy $\widehat{\sigma\tau} = \hat{\tau}\hat{\sigma}$.

These definitions lead to the following commutative diagram. For typographical reasons we identify each $[n]$ with its image $\Delta^n$ under the Yoneda embedding, and write $[-]$ for a generic such object of $\Delta$.

$(2.3.10)$
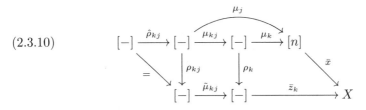

For each face operator $\mu \colon [m] \to [n]$, we let $\langle \mu \rangle \in |\Delta^n|$ be the barycenter of the face of $\Delta^n$ represented by $\mu$. As recalled in Example 2.2.4, this is the point with barycentric coordinates $(t_0, \ldots, t_n)$, where $t_i = 1/(m+1)$ if $i$ is in the image of $\mu$, and $t_i = 0$ otherwise. We shall make use of the barycenters of the composite face operators $\mu_j \hat{\rho}_{kj}$ for $0 \leq j \leq k \leq q$.

The map $h_x \colon |B(\Delta^n)| \to |\Delta^n|$ is defined in [FP67, (4)] by the explicit formula

$$h_x(u) = \sum_{0 \leq j \leq q} t_j (1 - t_n - \cdots - t_{j+1}) \langle \mu_j \hat{\rho}_{jj} \rangle + \sum_{0 \leq j < k \leq q} t_j t_k \langle \mu_j \hat{\rho}_{kj} \rangle,$$

due to Puppe (cf. [FP90, p. 222]). Note that $\rho_{jj} = \rho_j$. As a notational convention, we let

(2.3.11) $$\rho_{kj} = \rho_j \qquad \text{when } k \leq j.$$

Using that $(1 - t_n - \cdots - t_{j+1}) = \sum_{k=0}^{j} t_k$ for barycentric coordinates, we can rewrite the formula above as

$$h_x(u) = \sum_{0 \leq j \leq q} \sum_{0 \leq k \leq j} t_j t_k \langle \mu_j \hat{\rho}_{kj} \rangle + \sum_{0 \leq j < k \leq q} t_j t_k \langle \mu_j \hat{\rho}_{kj} \rangle,$$

and then collect the terms as

(2.3.12) $$h_x(u) = \sum_{j,k=0}^{q} t_j t_k \langle \mu_j \hat{\rho}_{kj} \rangle.$$

This is the formula we shall prefer to work with.

The sum expression (2.3.8) for $u$ is not unique, because the $q$-simplex in question may be degenerate and $(t_0, \ldots, t_q) \in |\Delta^q|$ might not be an interior point, but one can pass from any one such sum presentation of $u$ to any other by a finite chain of the following two operations: (a) deleting a term $t_i(\mu_i)$ when $t_i = 0$, and (b) combining two terms $t_i(\mu_i) + t_{i+1}(\mu_{i+1})$ to $(t_i + t_{i+1})(\mu_i)$ when $\mu_i = \mu_{i+1}$.

Fritsch and Puppe point out that each of these operations leave the point $h_x(u)$ defined by the expression (2.3.12) unchanged. Hence $h_x \colon |B(\Delta^n)| \to |\Delta^n|$ is a well-defined, continuous map. Furthermore, they verify that these maps satisfy Propositions 2.3.5 and 2.3.7, hence define a homeomorphism $h_X \colon |Sd(X)| \cong |X|$. That completes the proof of their theorem. We do not reproduce the rest of their arguments, because we shall need to generalize them for the proof of Theorem 2.3.2. $\square$

*Remark 2.3.13.* Note that $h_x(u)$ is in general a quadratic, rather than linear, expression in the barycentric coordinates of $u$. No linear expression can be found such that Proposition 2.3.5 holds, as explained in [FP67, p. 512]. In particular, the construction given in [Ba56, §3] does not work.

*Remark 2.3.14.* If each face $\mu_k^*(x)$ is already non-degenerate, as is the case when $x$ is non-degenerate and $X$ is non-singular, then each $\rho_k$ is the identity, so each $\rho_{kj}$ and $\hat{\rho}_{kj}$ is the identity, and $\langle \mu_j \hat{\rho}_{kj} \rangle = \langle \mu_j \rangle$. Hence in this case

$$h_x(u) = \sum_{j,k=0}^{q} t_j t_k \langle \mu_j \rangle = \sum_{j=0}^{q} t_j \langle \mu_j \rangle,$$

and $h_x \colon |B(\Delta^n)| \to |\Delta^n|$ equals the canonical homeomorphism $h_n$ of Example 2.2.4. Thus, for non-singular $X$ the Fritsch–Puppe homeomorphism $h_X \colon |Sd(X)| \cong |X|$ specializes to the homeomorphism we first constructed for non-singular simplicial sets.

We now turn to the case of subdividing a map $f \colon X \to Y$ of simplicial sets. To see how the homeomorphisms $h_X$ and $h_Y$ from Theorem 2.3.1 can fail to be natural, suppose that $X$ and $Y$ are non-singular, consider a non-degenerate simplex $x \in X_n$, and factor its image $f(x)$ as $\rho^*(y)$ for a degeneracy operator $\rho \colon [n] \to [p]$ and a non-degenerate simplex $y \in Y_p$. Then in the (non-commutative) square

$$
\begin{array}{ccc}
|B(\Delta^n)| & \xrightarrow{\ h_x\ } & |\Delta^n| \\
{\scriptstyle |B(\rho)|}\Big\downarrow & & \Big\downarrow{\scriptstyle |\rho|} \\
|B(\Delta^p)| & \xrightarrow{\ h_y\ } & |\Delta^p|
\end{array}
$$

the map $h_x = h_n$ takes the 0-cell $(\iota_n)$ of $|B(\Delta^n)|$ to the barycenter $\beta_n = \langle \iota_n \rangle$ in $|\Delta^n|$, while $|B(\rho)|$ takes $(\iota_n)$ to the 0-cell $(\iota_p)$ in $|B(\Delta^p)|$, which maps by $h_y = h_p$ to the barycenter $\beta_p = \langle \iota_p \rangle$ in $|\Delta^p|$. The problem is that the affine linear map $|\rho|$ does not in general take $\beta_n$ to $\beta_p$. Hence the displayed square does not in general commute. By naturality, using the square (2.3.6) for $x \in X_n$ and $y \in Y_p$, the same problem applies to the square in Theorem 2.3.2, if we try to use $h_X$ in place of the (yet to be specified) homeomorphism $h^f$.

Retaining the assumption that $Y$ is non-singular, our solution is to leave $h_Y$ unchanged, and to replace the maps $h_x$ by maps $h_x^f \colon |B(\Delta^n)| \to |\Delta^n|$ that take the 0-cell $(\iota_n)$ to an interior point of $|\Delta^n|$ that is mapped by $|\rho|$ to the barycenter of $|\Delta^p|$. A suitable such point is the pseudo-barycenter $\beta_n^\rho \in |\Delta^n|$, which depends not only on the simplex $\Delta^n$, but also on the surjective map $\rho \colon \Delta^n \to \Delta^p$.

**Definition 2.3.15.** Let $\rho \colon [n] \to [p]$ be a degeneracy operator. For each $i \in [p]$ let $\beta_{(i)} \in |\Delta^n|$ be the barycenter of the preimage $|\rho|^{-1}(i) \subset |\Delta^n|$ of the vertex $(i) \in |\Delta^p|$. This preimage is the face of $|\Delta^n|$ spanned by the vertices in $\rho^{-1}(i) \subset [n]$. Let the **pseudo-barycenter** $\beta_n^\rho \in |\Delta^n|$, **with respect to** $\rho$, be the barycenter of the affine $p$-simplex $s_p \subset |\Delta^n|$ spanned by the points $\beta_{(0)}, \dots, \beta_{(p)}$.

Then $|\rho| \colon |\Delta^n| \to |\Delta^p|$ maps $s_p$ homeomorphically onto $|\Delta^p|$, and takes the pseudo-barycenter $\beta_n^\rho$ to the barycenter $\beta_p$ of $|\Delta^p|$. In particular, $\beta_n^\rho$ is an interior point of $|\Delta^n|$. For an equivalent definition, let $n_j$ be the cardinality of $\rho^{-1}\rho(j) = \{k \in [n] \mid \rho(k) = \rho(j)\}$, for each $j \in [n]$. Then $\beta_n^\rho = (t_0, \dots, t_n)$ with $t_j = 1/n_j(p+1)$.

**Definition 2.3.16.** Let $f \colon X \to Y$ be a map of simplicial sets, $x \in X_n$ an $n$-simplex and $\mu \colon [m] \to [n]$ a face operator. Write $\mu^* f(x) = f(\mu^*(x)) \in Y_m$ as $\gamma^*(y)$ where $\gamma \colon [m] \to [r]$ is a degeneracy operator and $y \in Y_r$ is non-degenerate. Define the **pseudo-barycenter of the face** $\mu$ **of** $x$, **with respect to** $f$, to be the point

$$\langle \mu \rangle_x^f = |\mu|(\beta_m^\gamma)$$

in $|\Delta^n|$. This is the image under $|\mu|$ of the pseudo-barycenter $\beta_m^\gamma$ of $|\Delta^m|$ with respect to $\gamma$. In the geometric realization of the following commutative diagram of simplicial sets,

$$
\begin{array}{ccc}
\Delta^m \xrightarrow{\ \mu\ } \Delta^n \xrightarrow{\ \bar{x}\ } X \\
\gamma \downarrow \qquad\qquad\qquad \downarrow f \\
\Delta^r \xrightarrow{\qquad \bar{y} \qquad} Y
\end{array}
$$

the pseudo-barycenter $\beta_m^\gamma \in |\Delta^m|$ maps to the barycenter $\beta_r \in |\Delta^r|$ and to $\langle \mu \rangle_x^f \in |\Delta^n|$.

*Proof of Theorem 2.3.2, the non-singular case.* Suppose that $f\colon X \to Y$ is a map of finite non-singular simplicial sets. For each non-degenerate simplex $x \in X$, of dimension $n$, say, factor $f(x) \in Y$ as $\rho^*(y)$, for a degeneracy operator $\rho\colon [n] \to [p]$ and a non-degenerate $p$-simplex $y \in Y$. For each face operator $\mu\colon [m] \to [n]$, factor $\mu^*(\rho) = \rho\mu$ as $\gamma^*(\nu) = \nu\gamma$, for a degeneracy operator $\gamma\colon [m] \to [r]$ and a face operator $\nu\colon [r] \to [p]$.

$$
\begin{array}{ccc}
\Delta^m \xrightarrow{\ \mu\ } \Delta^n \xrightarrow{\ \bar{x}\ } X \\
\gamma \downarrow \qquad \rho \downarrow \qquad\quad \downarrow f \\
\Delta^r \xrightarrow{\ \nu\ } \Delta^p \xrightarrow{\ \bar{y}\ } Y
\end{array}
$$

Then $\nu^*(y)$ is non-degenerate, because $Y$ is non-singular, and $\gamma^*(\nu^*(y))$ is the Eilenberg–Zilber factorization of $f(\mu^*(x))$. Define a map

$$h_x^f\colon |Sd(\Delta^n)| \to |\Delta^n|$$

$$(\mu) \mapsto \langle \mu \rangle_x^f$$

by taking the point corresponding to the 0-simplex $(\mu)$ of $Sd(\Delta^n)$ to the pseudo-barycenter $\langle \mu \rangle_x^f = |\mu|(\beta_m^\gamma)$ of the face $\mu$, with respect to $\gamma$, and extending affine linearly on each simplex of $Sd(\Delta^n)$. In particular, $h_x^f$ takes $(\iota_n)$ to the pseudo-barycenter $\beta_n^\rho$, which is an interior point of $|\Delta^n|$. Furthermore, $h_x^f$ takes the boundary $|Sd(\partial \Delta^n)|$ to $|\partial \Delta^n|$, and similarly for each face of $\Delta^n$.

Let $h_Y\colon |Sd(Y)| \to |Y|$ be the homeomorphism from Theorem 2.3.1. We claim that there is a (piecewise-linear) homeomorphism $h^f\colon |Sd(X)| \to |X|$ with $|f| \circ h^f = h_Y \circ |Sd(f)|$, such that for each non-degenerate simplex $x$ of $X$ the square

$$
\begin{array}{ccc}
|Sd(\Delta^n)| & \xrightarrow{\ h_x^f\ } & |\Delta^n| \\
{\scriptstyle |Sd(\bar{x})|} \downarrow & & \downarrow {\scriptstyle |\bar{x}|} \\
|Sd(X)| & \xrightarrow{\ h^f\ } & |X|
\end{array}
$$

commutes, where $\bar{x}\colon \Delta^n \to X$ is the representing map of $x$, as usual. By induction, we can write $X = \Delta^n \cup_{\partial\Delta^n} X'$, with $\partial\Delta^n \to X'$ a cofibration, and assume that $h^{f|\Delta^n} = h^f_x$, $h^{f|\partial\Delta^n}$ and $h^{f|X'}$ exist, with the characterizing property just given.

$$
\begin{array}{ccc}
|Sd(\Delta^n)| & \longleftarrow |Sd(\partial\Delta^n)| \longrightarrow |Sd(X')| \\
\downarrow{\scriptstyle h^f_x} & \downarrow{\scriptstyle h^{f|\partial\Delta^n}} \qquad \downarrow{\scriptstyle h^{f|X'}} \\
|\Delta^n| & \longleftarrow |\partial\Delta^n| \longrightarrow |X'|
\end{array}
$$

To define $h^f$ as the pushout of these three maps, we need to know that the left hand square commutes. But this is clear, because $h^f_x$ is constructed to be natural for face operators.

By induction, $h^{f|\partial\Delta^n}$ is a homeomorphism, and $h^f_x$ is the cone on this map, via the identifications $|Sd(\Delta^n)| \cong \mathrm{cone}\,|Sd(\partial\Delta^n)|$ and $|\Delta^n| \cong \mathrm{cone}\,|\partial\Delta^n|$, with cone points $(\iota_n)$ and $\beta^\rho_n$, respectively. Hence $h^f_x$ is also a homeomorphism. It follows by induction that also $h^f$ is a homeomorphism. $\square$

The remainder of this section is concerned with the proof of Theorem 2.3.2, in the case when $X$ is not necessarily non-singular.

**Lemma 2.3.17.** *Let* $\alpha\colon [n] \to [p]$ *be a morphism in* $\Delta$ *and* $\mu\colon [m] \to [n]$ *a face operator such that* $\alpha\mu\colon [m] \to [p]$ *is also a face operator. Then for each pair of simplices* $x \in X_n$, $y \in X_p$ *with* $x = \alpha^*(y)$ *we have*

$$
|\alpha|\langle\mu\rangle^f_x = \langle\alpha\mu\rangle^f_y
$$

*in* $|\Delta^p|$.

*Proof.* We factor $\mu^* f(x) = (\alpha\mu)^* f(y)$ as $\gamma^*(w)$ for a degeneracy operator $\gamma\colon [m] \to [r]$ and a non-degenerate simplex $w \in Y_r$.

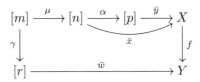

Then we can compute

$$
|\alpha|\langle\mu\rangle^f_x = |\alpha||\mu|(\beta^\gamma_m) = |\alpha\mu|(\beta^\gamma_m) = \langle\alpha\mu\rangle^f_y .
$$

$\square$

*Proof of Theorem 2.3.2, the general case.* Preserving the notations from diagram (2.3.10), and the convention $\rho_{kj} = \rho_j$ for $k \le j$ from (2.3.11), we define $h^f_x\colon |B(\Delta^n)| \to |\Delta^n|$ at a point $u = \sum^q_{k=0} t_k(\mu_k)$ by the explicit formula

$$
(2.3.18) \qquad\qquad h^f_x(u) = \sum^q_{j,k=0} t_j t_k \langle\mu_j\hat\rho_{kj}\rangle^f_x .
$$

Compared to formula (2.3.12), the barycenters $\langle \mu \rangle$ for faces of simplices of $X$ are replaced by the corresponding pseudo-barycenters with respect to the map $f$. In the style of [FP67], without our notational convention, the formula can be written as

$$h_x^f(u) = \sum_{0 \le j \le q} t_j(1 - t_n - \cdots - t_{j+1})\langle \mu_j \hat{\rho}_{jj} \rangle_x^f + \sum_{0 \le j < k \le q} t_j t_k \langle \mu_j \hat{\rho}_{kj} \rangle_x^f .$$

**Lemma 2.3.19.** *The formula (2.3.18) above defines a well-defined, continuous map*

$$h_x^f \colon |B(\Delta^n)| \to |\Delta^n| .$$

*Proof.* If some $t_i = 0$, the term $t_i(\mu_i)$ may be deleted from the sum (2.3.8) without altering $u$. Then the terms in (2.3.18) for $j = i$ or $k = i$ also disappear, but all these terms were 0 due to the vanishing coefficient $t_j t_k$, so $h_x^f(u)$ is unchanged.

If some $\mu_i = \mu_{i+1}$, the sum of terms $t_i(\mu_i) + t_{i+1}(\mu_{i+1})$ in (2.3.8) may be replaced with a single term $(t_i + t_{i+1})(\mu_i)$. This affects the terms in (2.3.18) for $j$ or $k$ equal to $i$ or $i+1$. But $\rho_{ki} = \rho_{k,i+1}$ for $k \ge i+1$, $\rho_{ij} = \rho_{i+1,j}$ for $j \le i$ and $\rho_i = \rho_{i+1,i} = \rho_{i+1}$, so by a little calculation the change does not affect the value of the sum $h_x^f(u)$. $\square$

Following the proof of [FP67, Satz], we claim that $h_x^f$ satisfies the following two propositions.

**Proposition 2.3.20.** *For every morphism* $\alpha \colon [n] \to [p]$ *in* $\Delta$, *and pair of simplices* $x \in X_n$, $y \in X_p$ *with* $\alpha^*(y) = x$, *the square*

$$
\begin{array}{ccc}
|B(\Delta^n)| & \xrightarrow{\; h_x^f \;} & |\Delta^n| \\
{\scriptstyle |B(\alpha)|} \big\downarrow & & \big\downarrow {\scriptstyle |\alpha|} \\
|B(\Delta^p)| & \xrightarrow{\; h_y^f \;} & |\Delta^p|
\end{array}
$$

*commutes.*

*Proof.* The proof closely follows that of [FP67, (A)], using Lemma 2.3.17. Let $u \in |B(\Delta^n)|$ be expressed as in (2.3.8), and let

$$B(\alpha)(\mu_0 \le \cdots \le \mu_q) = (\nu_0 \le \cdots \le \nu_q).$$

This means that $\alpha \mu_k = \nu_k \tau_k$ for unique face operators $\nu_k$ with target $[p]$, and degeneracy operators $\tau_k$, for $0 \le k \le q$. By definition, $|B(\alpha)|(u) = \sum_{k=0}^q t_k(\nu_k)$.

We now follow the definitions leading to the diagram (2.3.10), but applied to the $q$-simplex $(\nu_0 \le \cdots \le \nu_q)$ and $y \in X_p$. For each $0 \le j \le k \le q$ there is a unique face operator $\nu_{kj}$ such that $\nu_k \nu_{kj} = \nu_j$. Then $\tau_k \mu_{kj} = \nu_{kj} \tau_j$, because $\nu_k$ is injective.

For each $0 \leq k \leq q$ the face $\nu_k^*(y)$ factors uniquely as $\sigma_k^*(z_k)$ for a non-degenerate simplex $z_k$ in $X$ and a degeneracy operator $\sigma_k$. Then $\rho_k = \sigma_k \tau_k$ and $z_k$ is the same as in (2.3.10), by the uniqueness of the factorization for $\mu_k^*(x)$.

For each $0 \leq j \leq k \leq q$ we uniquely factor the composite morphism $\sigma_k \nu_{kj}$ in $\Delta$ as a degeneracy operator $\sigma_{kj}$ followed by a face operator $\tilde{\nu}_{kj}$. Then $\rho_{kj} = \sigma_{kj}\tau_j$ and $\tilde{\mu}_{kj} = \tilde{\nu}_{kj}$, by the uniqueness of the factorization. We obtain the following commutative diagram:

(2.3.21)

Thus the composite

(2.3.22) $$\alpha \mu_j \hat{\rho}_{kj} = \nu_j \tau_j \hat{\tau}_j \hat{\sigma}_{kj} = \nu_j \hat{\sigma}_{kj}$$

is a face operator. We let $\sigma_{kj} = \sigma_j$ when $k \leq j$. Then by Lemma 2.3.17 and (2.3.22) we get

$$|\alpha| h_x^f(u) = \sum_{j,k=0}^{q} t_j t_k |\alpha| \langle \mu_j \hat{\rho}_{kj} \rangle_x^f$$

$$= \sum_{j,k=0}^{q} t_j t_k \langle \alpha \mu_j \hat{\rho}_{kj} \rangle_y^f = \sum_{j,k=0}^{q} t_j t_k \langle \nu_j \hat{\sigma}_{kj} \rangle_y^f = h_y^f |B(\alpha)|(u).$$

$\square$

**Proposition 2.3.23.** *For each non-degenerate simplex $x \in X_n$, the map $h_x^f$ takes the interior of $|B(\Delta^n)|$ bijectively onto the interior of $|\Delta^n|$.*

*Proof.* Keeping in mind that each pseudo-barycenter $\beta_n^\gamma$ is an interior point of $|\Delta^n|$, the proof closely follows that of [FP67, (B)]. Each point $u \in |B(\Delta^n)|$ can be represented in the form

$$u = \sum_{k=0}^{n} t_k(\mu_k)$$

where $\mu_k \colon [k] \to [n]$ for each $0 \leq k \leq n$, by possibly letting some $t_k = 0$. There is a permutation $\varphi$ of $[n]$ such that the image of $\mu_k$ is $\varphi([k])$, for each $k$. Since $x$ is non-degenerate, $\rho_n = \iota_n$, and it follows that $\rho_{nj} = \iota_j$ for each $0 \leq j \leq n$.

**Lemma 2.3.24.** *For $x \in X_n$ non-degenerate, $h_x^f$ maps the interior of $|B(\Delta^n)|$ to the interior of $|\Delta^n|$.*

*Proof.* By our notational convention, $\rho_{kn} = \iota_n$ for $0 \leq k \leq n$. The $i$-th barycentric coordinates in $|\Delta^n|$, indicated by a subscript $i$, satisfy

$$h_x^f(u)_i \geq t_n \langle \iota_n \rangle_{x,i}^f$$

(consider only the terms with $j = n$ in (2.3.18)). When $u$ is interior we have $t_n > 0$, and the pseudo-barycenter $\langle \iota_n \rangle_x^f$ is interior, so each of these barycentric coordinates is positive. Hence $h_x^f(u)$ is interior. $\square$

To show that $h_x^f$ is injective on the interior of $|B(\Delta^n)|$, consider a second point $u' = \sum_{k=0}^n t_k'(\mu_k')$ in that interior. We decorate the constructions leading to (2.3.10), but based on $u'$, with a prime.

**Lemma 2.3.25.** *Suppose that $h_x^f(u) = h_x^f(u')$, with $u$ and $u'$ interior points in $|B(\Delta^n)|$. Then $t_j = t_j'$, and $\mu_j = \mu_j'$ if $t_j > 0$, for each $0 \leq j \leq n$. Hence $u = u'$.*

*Proof.* We prove the statement "$t_j = t_j'$, and $\mu_j = \mu_j'$ if $t_j > 0$" by descending induction on $0 \leq j \leq n$. To start the induction, note from Definition 2.3.16 that the $i$-th barycentric coordinate satisfies $\langle \mu_j \hat{\rho}_{kj} \rangle_{x,i}^f = 0$ for $i$ not in the image $\varphi([j])$ of $\mu_j$. Then the $\varphi(n)$-th barycentric coordinates satisfy

$$t_n \langle \iota_n \rangle_{x,\varphi(n)}^f = h_x^f(u)_{\varphi(n)} = h_x^f(u')_{\varphi(n)} \geq t_n' \langle \iota_n \rangle_{x,\varphi(n)}^f$$

(consider the terms with $j = n$ in $h_x^f(u')$). Since $\langle \iota_n \rangle_x^f$ is interior, this implies $t_n \geq t_n'$. By symmetry $t_n = t_n'$, and we already know that $\mu_n = \mu_n' = \iota_n$.

For the inductive step, let $0 \leq \ell < n$ and assume that the statement has been proved for all $j > \ell$. We must prove the statement for $j = \ell$. Consider the point

$$T = \sum_{j=0}^{\ell} \sum_{k=0}^{n} t_j t_k \langle \mu_j \hat{\rho}_{kj} \rangle_x^f$$

obtained from $h_x^f(u)$ by deleting the terms with $j > \ell$. By the inductive hypothesis, $T$ equals the corresponding point $T'$ derived from $u'$.

If $\mu_\ell \neq \mu_\ell'$, then there is an $i \in [n]$ that is in the image of $\mu_\ell$, but not in the image of $\mu_\ell'$. Then the $i$-th barycentric coordinate of $T = T'$ satisfies

$$t_\ell t_n \langle \mu_\ell \rangle_{x,i}^f \leq T_i = T_i' = 0$$

(consider only the term in $T$ with $j = \ell$ and $k = n$). Since $\langle \mu_\ell \rangle_x^f$ is interior in the $\mu_\ell$-th face, its $i$-th coordinate is positive, and $t_n > 0$ because $u$ is interior, so under this hypothesis we must have $t_\ell = 0$. By symmetry, also $t_\ell' = 0$.

If on the other hand $\mu_\ell = \mu'_\ell$, consider the point

$$S = \sum_{k=0}^{n} t_k \langle \mu_\ell \hat{\rho}_{k\ell} \rangle_x^f$$

and the corresponding point $S'$. (There is a typographical error at this step in [FP90, p. 203].) These are equal by the inductive hypothesis. With $i = \varphi(\ell)$ we have that $i$ is in the image of $\mu_\ell$, but not in the image of $\mu_j$ for $j < \ell$, so the $i$-th barycentric coordinates satisfy

$$t_\ell S_i = T_i = T'_i \geq t'_\ell S'_i = t'_\ell S_i .$$

Furthermore, $S_i \geq t_n \langle \mu_\ell \rangle_{x,i}^f > 0$, because this pseudo-barycenter is interior in the $\mu_\ell$-th face, and $u$ is interior. Therefore we must have $t_\ell \geq t'_\ell$, and by symmetry, $t_\ell = t'_\ell$. This establishes the inductive step, and completes the proof. $\square$

We now return to the proof of Proposition 2.3.23. By Lemma 2.3.25 the map $h_x^f$ is injective when restricted to the interior of $|B(\Delta^n)|$, so by the theorem of invariance of domain [HW41, Thm. VI.9], this restricted image is open in the interior of $|\Delta^n|$. By Proposition 2.3.20 the boundary of $|B(\Delta^n)|$ is mapped into the boundary of $|\Delta^n|$, and $h_x^f$ is a closed map, so the image of $h_x^f$ on the interior is an open, closed and non-empty subset of the interior of $|\Delta^n|$, hence equals all of this (connected) interior. Thus $h_x^f$ is also surjective as a map of interiors. This concludes the proof of Proposition 2.3.23. $\square$

By Proposition 2.3.20 the maps $h_x^f$ assemble to a map

$$h^f = \operatorname*{colim}_{\text{simp}(X)} \left( ([n], x) \mapsto h_x^f \right) \colon |Sd(X)| \to |X|$$

and by Proposition 2.3.23 this map is a homeomorphism.

It remains to verify that the desired quasi-naturality property holds, i.e., that $|f| \circ h^f = h_Y \circ |Sd(f)|$. It suffices to check this after composing with $|Sd(\bar{x})|$ on the right, for every (non-degenerate) simplex $x$ of $X$, because the images of the latter maps cover $|Sd(X)|$. We factor $f(x)$ as $\gamma^*(y)$ for a degeneracy operator $\gamma \colon [n] \to [p]$ and a non-degenerate simplex $y \in Y_p$. Then in the cubical diagram

(2.3.26)

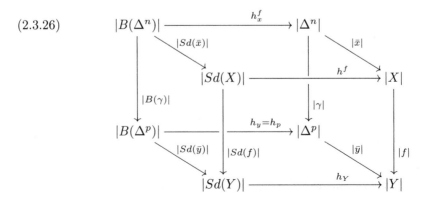

the left and right hand faces are obtained by applying the normal subdivision and geometric realization functors to the relation $f\bar{x} = \bar{y}\gamma$, and therefore commute. The top and bottom faces commute by the definition of the maps $h^f$ and $h_Y$ as the colimits of the maps $h_x^f$ and $h_y$, respectively. We note as in Remark 2.3.14 that $h_y = h_p$ is the canonical map from Example 2.2.4, because $Y$ is non-singular.

Thus to show that the front face of the cube commutes, it suffices to show that the back face commutes for each simplex $x$ of $X$. The following proposition will therefore complete the proof of Theorem 2.3.2. $\square$

**Proposition 2.3.27.** *Let $f\colon X \to Y$ be a map of simplicial sets, with $Y$ non-singular, and let $f(x) = \gamma^*(y)$ for $x \in X_n$, $\gamma\colon [n] \to [p]$ a degeneracy operator and $y \in Y_p$ non-degenerate. Then the square*

$$
\begin{array}{ccc}
|B(\Delta^n)| & \xrightarrow{\ h_x^f\ } & |\Delta^n| \\
{\scriptstyle |B(\gamma)|}\big\downarrow & & \big\downarrow{\scriptstyle |\gamma|} \\
|B(\Delta^p)| & \xrightarrow{\ h_p\ } & |\Delta^p|
\end{array}
$$

*commutes.*

*Proof.* We keep the notation of the sum formula (2.3.8) for $u \in |B(\Delta^n)|$, the diagram (2.3.10) and the explicit formula (2.3.18) for $h_x^f(u) \in |\Delta^n|$. Thus

$$
|\gamma| h_x^f(u) = \sum_{j,k=0}^{q} t_j t_k |\gamma| \langle \mu_j \hat{\rho}_{kj} \rangle_x^f \, .
$$

We uniquely factor $\gamma\mu_j = \lambda_j \pi_j$ for each $0 \le j \le q$, where $\lambda_j$ is a face operator with target $[p]$ and $\pi_j$ is a degeneracy operator. Then $B(\gamma)(\mu_0 \le \cdots \le \mu_q) = (\lambda_0 \le \cdots \le \lambda_q)$ and

$$
h_p |B(\gamma)|(u) = \sum_{j=0}^{q} t_j \langle \lambda_j \rangle \, .
$$

It will therefore suffice to show that

$$
|\gamma| \langle \mu_j \hat{\rho}_{kj} \rangle_x^f = \langle \lambda_j \rangle
$$

in $|\Delta^p|$, for all $0 \le j \le k \le q$. In particular, the left hand side is independent of $k$.

Fix a pair $j \le k$, and factor $\pi_j \hat{\rho}_{kj} = \nu\tau$, where $\nu$ is a face operator and $\tau\colon [m] \to [r]$ is a degeneracy operator. We obtain the following commutative diagram:

$$
(2.3.28) \qquad
\begin{array}{ccccccc}
[m] & \xrightarrow{\ \hat{\rho}_{kj}\ } & [-] & \xrightarrow{\ \mu_j\ } & [n] & \xrightarrow{\ \bar{x}\ } & X \\
{\scriptstyle \tau}\big\downarrow & & {\scriptstyle \pi_j}\big\downarrow & & {\scriptstyle \gamma}\big\downarrow & & \big\downarrow{\scriptstyle f} \\
[r] & \xrightarrow{\ \nu\ } & [-] & \xrightarrow{\ \lambda_j\ } & [p] & \xrightarrow{\ \bar{y}\ } & Y \, .
\end{array}
$$

Here the face $w = (\lambda_j \nu)^*(y) \in Y_r$ of $y$ is non-degenerate, because $y$ is non-degenerate and $Y$ is non-singular, so $(\mu_j \hat\rho_{kj})^* f(x) = \tau^*(w)$ is the factorization to be used in defining the pseudo-barycenter $\langle \mu_j \hat\rho_{kj} \rangle_x^f = |\mu_j \hat\rho_{kj}|(\beta_m^\tau)$ of the face $\mu_j \hat\rho_{kj}$ of $x$ with respect to $f$.

The pseudo-barycenter $\beta_m^\tau$ is chosen to map under $|\tau|$ to the (ordinary) barycenter $\beta_r$ of $|\Delta^r|$, so

$$|\gamma| \langle \mu_j \hat\rho_{kj} \rangle_x^f = |\gamma \mu_j \hat\rho_{kj}|(\beta_m^\tau) = |\lambda_j \nu \tau|(\beta_m^\tau) = |\lambda_j \nu|(\beta_r) = \langle \lambda_j \nu \rangle$$

in $|\Delta^p|$. In fact, we claim that for $Y$ non-singular the image of $\gamma \mu_j \hat\rho_{kj}$ in $[p]$ equals the image of $\gamma \mu_j$, so $\lambda_j \nu = \lambda_j$ (and $\nu$ is the identity morphism). To prove this claim, we use the following lemma.

**Lemma 2.3.29.** *When $Y$ is non-singular, the composite operation $\gamma \mu_j$ extends over the surjection $\rho_{kj}$, as $\gamma \mu_j = \alpha \rho_{kj}$ for a unique morphism $\alpha$ in $\Delta$.*

*Proof.* Suppose that two elements in the source of $\mu_j$ have the same image under $\rho_{kj}$. Regarded as 0-simplices, they then have the same image under $\bar x \mu_j$ by (2.3.10), and thus have the same image under $\bar y \gamma \mu_j$ by (2.3.28). But $Y$ is non-singular, so $\bar y$ is a cofibration, and the two elements also have the same image under $\gamma \mu_j$. Hence there is a factorization of $\gamma \mu_j$ over $\rho_{kj}$, as asserted. It is unique because $\rho_{kj}$ is surjective. $\square$

Thus the image of the face operator $\lambda_j \nu$ equals the image of $\gamma \mu_j \hat\rho_{kj}$, because $\tau$ is surjective and (2.3.28) commutes. The latter image equals the image of $\alpha \rho_{kj} \hat\rho_{kj} = \alpha$ by Lemma 2.3.29, which in turn equals the image of $\alpha \rho_{kj} = \gamma \mu_j$ because $\rho_{kj}$ is surjective. By (2.3.28) this equals the image of the face operator $\lambda_j$, and so $\lambda_j \nu = \lambda_j$. This concludes the proof of Proposition 2.3.27. $\square$

## 2.4. THE REDUCED MAPPING CYLINDER

To each map $f \colon X \to Y$ of simplicial sets, we shall naturally associate a simplicial set $M(f)$ called the reduced mapping cylinder of $f$. There is a natural cylinder reduction map $T(f) = X \times \Delta^1 \cup_X Y \to M(f)$ from the ordinary mapping cylinder to the reduced one. There is also a natural cofibration $M(f) \to \mathrm{cone}(X) \times Y$. These two maps are compatible with front inclusions of $X$, back inclusions of $Y$, and cylinder projections to $Y$.

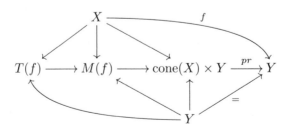

Like the ordinary mapping cylinder $T(f)$, the reduced mapping cylinder $M(f)$ and the cone-product construction $\mathrm{cone}(X) \times Y$ each come in a forward and a backward version. In the forward versions, there are 1-simplices going from the image of the front inclusion to the image of the back inclusion, while in the backward versions these 1-simplices are oriented the other way. The backward (reduced or ordinary) mapping cylinder of $f$ equals the simplicial opposite of the forward (reduced or ordinary) mapping cylinder of $f^{op}$.

Starting with Proposition 2.4.16, and for the remainder of this book, we will only make use of the backward version of the reduced mapping cylinder. So outside of the present section, the notation $M(f)$ will always refer to the backward reduced mapping cylinder. The asymmetry enters in Lemma 2.4.21, where we emphasize the faces of a given simplex $x$, rather than its cofaces (containing $x$ as a face).

When $f = Sd(g) \colon Sd(X) \to Sd(Y)$, for a map $g \colon X \to Y$ of ordered simplicial complexes, the backward reduced mapping cylinder $M(Sd(g))$ equals Cohen's mapping cylinder $C_g$, as defined in [Co67, §4], in the subdivided form that contains $Sd(X)$ and $Sd(Y)$ at the front and back ends, respectively. This is clear from Lemma 2.4.12. See also Remark 4.3.2.

We say that a map $f$ has simple cylinder reduction when the map $T(f) \to M(f)$ of backward mapping cylinders is simple. We show in Lemma 2.4.21 and Corollary 2.5.7 that $f$ has simple cylinder reduction when $f = Sd(g)$ for a map $g \colon X \to Y$ of non-singular simplicial sets, and more generally, when $f = B(g)$ for $g \colon X \to Y$ a map of op-regular simplicial sets. See also Lemma 2.7.4. In these cases, then, $M(f)$ is really a reduced version of $T(f)$.

Let $f \colon X \to Y$ be a map of simplicial sets. The definition of the reduced mapping cylinder $M(f)$ will be parallel to that of the normal subdivision functor. First we consider the "local case" of a map of simplices $N\varphi \colon \Delta^n \to \Delta^m$, and define $M(N\varphi)$ in a combinatorial fashion. Thereafter we glue these building blocks together by means of a colimit construction. For $M(f)$ to contain $Y$ as a simplicial subset, the simplices of $Y$ that are not in the image of $f$ must be dealt with separately. We shall arrange this by means of the following convention on empty $(-1)$-simplices, which will be in effect throughout this section.

**Definition 2.4.1.** Let $[-1]$ be the empty set with its unique total ordering, and let $\Delta^{-1} = N[-1]$ be the empty simplicial set. Let $\Delta_\eta$ be the skeleton category of finite ordinals, with objects the $[n]$ for $n \geq -1$ and morphisms the order-preserving functions. For each simplicial set $X$ we interpret the unique map $\bar{\eta} \colon \Delta^{-1} \to X$ as the representing map of a unique $(-1)$-simplex $\eta \in X_{-1}$. Then each simplicial set $X$ extends uniquely to a contravariant functor from $\Delta_\eta$ to sets, whose value at $[-1]$ is the one-point set $X_{-1} = \{\eta\}$, and conversely, all contravariant functors from $\Delta_\eta$ to sets with value $\{\eta\}$ at $[-1]$ arise in this way.

**Definition 2.4.2.** Let $\mathrm{simp}(f)$ be the category with objects the commutative squares

$$\begin{array}{ccc}
\Delta^n & \xrightarrow{\bar{x}} & X \\
{\scriptstyle\varphi}\downarrow & & \downarrow{\scriptstyle f} \\
\Delta^m & \xrightarrow{\bar{y}} & Y
\end{array}$$

with $n, m \geq -1$, briefly denoted $(\varphi\colon [n] \to [m], x, y)$. Here $\varphi$ is any order-preserving function, $x \in X_n$ and $y \in Y_m$ are simplices and $f(x) = \varphi^*(y)$. A morphism $(\alpha, \beta)$ from $(\varphi'\colon [n'] \to [m'], x', y')$ to $(\varphi, x, y)$ is a pair of order-preserving functions $\alpha\colon [n'] \to [n]$, $\beta\colon [m'] \to [m]$ such that $\alpha^*(x) = x'$, $\beta^*(y) = y'$ and $\varphi\alpha = \beta\varphi'$. In other words, $\mathrm{simp}(f) = \Upsilon/f$ is the left fiber at $f\colon X \to Y$ of the Yoneda embedding $\Upsilon$ of the category of arrows in $\Delta_\eta$ into the category of arrows in simplicial sets.

In the following three definitions, we discuss both forward and backward cases, without making this explicit in the notation. The context should make it clear which case is at hand.

**Definition 2.4.3.** For each order-preserving function $\varphi\colon C \to D$ of partially ordered sets let $P(\varphi) = C \sqcup_\varphi D$ be the forward (resp. backward) partially ordered set obtained from the disjoint union of $C$ and $D$ by adjoining the relations $c < \varphi(c)$ (resp. $c > \varphi(c)$) for all $c \in C$.

Let $\pi\colon P(\varphi) \to [1] = \{0 < 1\}$ be the order-preserving function that takes each $c \in C$ to 0 (resp. to 1) and each $d \in D$ to 1 (resp. to 0).

**Definition 2.4.4.** For the map $N\varphi\colon \Delta^n \to \Delta^m$ of nerves induced by an order-preserving function $\varphi\colon [n] \to [m]$ let

$$M(N\varphi) = NP(\varphi) = N([n] \sqcup_\varphi [m])$$

be the nerve of the forward (resp. backward) partially ordered set.

The rule that takes the object $(\varphi, x, y)$ of $\mathrm{simp}(f)$ to $M(N\varphi)$, and the morphism $(\alpha, \beta)$ to the nerve of the order-preserving function $P(\alpha, \beta) = \alpha \sqcup \beta\colon P(\varphi') \to P(\varphi)$, defines a functor from $\mathrm{simp}(f)$ to simplicial sets. The **forward** (resp. **backward**) **reduced mapping cylinder** $M(f)$ of $f\colon X \to Y$ is defined to be the colimit

$$M(f) = \operatorname*{colim}_{\mathrm{simp}(f)} \left( (\varphi, x, y) \mapsto M(N\varphi) \right)$$

of this functor.

The backward partially ordered set $P(\varphi)$ has the partial ordering opposite to that of the forward partially ordered set $P(\varphi^{op})$. Hence the backward reduced mapping cylinder $M(f)$ of $f\colon X \to Y$ is the simplicial opposite of the forward reduced mapping cylinder of $f^{op}\colon X^{op} \to Y^{op}$.

We get a commutative diagram of partially ordered sets and order-preserving functions,

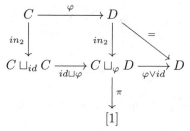

where the square is a pushout, and we can identify $C \sqcup_{id} C \cong C \times [1]$.

Passing to nerves, the square induces a cylinder reduction map from the pushout $T(N\varphi) = NC \times \Delta^1 \cup_{NC} ND$ to $M(N\varphi)$, where $NC \to NC \times \Delta^1$ has the form $id \times \delta^0$ (resp. $id \times \delta^1$). The folded map $\varphi \vee id$ induces a reduced cylinder projection $pr \colon M(N\varphi) \to ND$, and the composite map is the ordinary cylinder projection $T(N\varphi) \to ND$. The front inclusion $NC \to T(N\varphi)$ (derived from the inclusion $id \times \delta^1 \colon NC \to NC \times \Delta^1$, resp. $id \times \delta^0$) and the back inclusion $ND \to T(N\varphi)$ compose with the cylinder reduction map to give front and back inclusion maps to $M(N\varphi)$, respectively.

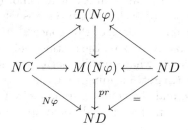

**Definition 2.4.5.** The diagram above is natural in $\varphi$, so we may form a colimit over $\mathrm{simp}(f)$ to obtain the following commutative diagram

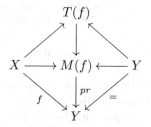

of simplicial sets. The maps from $X$ to $T(f)$ and $M(f)$ are the **front inclusions**, the maps from $Y$ to $T(f)$ and $M(f)$ are the **back inclusions**, and the maps from $T(f)$ and $M(f)$ to $Y$ are the **ordinary** and **reduced cylinder projections**, respectively. We call $T(f) \to M(f)$ the **cylinder reduction map**. There is also a **cylinder coordinate projection** $\pi \colon M(f) \to \Delta^1$, compatible with the one from $T(f)$.

*Remark 2.4.6.* The $q$-simplices of $M(f)$ can be explicitly described as the commutative diagrams

$$\begin{array}{ccc} [p] & \xrightarrow{\bar{x}} & X \\ \mu \downarrow & & \downarrow f \\ [q] & \xrightarrow{\bar{y}} & Y \end{array}$$

for $-1 \leq p \leq q$, where $\mu \colon [p] \subset [q]$ is a front face, $x \in X_p$ and $y \in Y_q$. We omit the proof.

We have already remarked in Section 2.1 that the ordinary cylinder projection $T(f) \to Y$ is always simple, because its geometric realization has point inverses that are cones, hence contractible. The reduced cylinder projection $M(f) \to Y$ shares this property. (We shall see in Lemma 2.4.11(c) that $M(f)$ is a finite simplicial set when $X$ and $Y$ are finite.) For the proof we shall use the following notion.

**Definition 2.4.7.** A map $f \colon X \to Y$ of finite simplicial sets is called a **homotopy equivalence over the target** if it has a section $s \colon Y \to X$ such that $sf$ is homotopic to the identity map on $X$, by a homotopy over $Y$. The homotopy provides a contraction of each point inverse, so such a map $f$ is simple. Technically there are several versions of this concept, in that one can allow for the existence of the homotopy, or even the section, only after geometric realization. The strongest notion is what we will refer to as a **simplicial homotopy equivalence over the target**, meaning that the section $s \colon Y \to X$ exists as a map of simplicial sets, and the composite map $sf \colon X \to X$ is homotopic to the identity on $X$ by a chain of simplicial homotopies over $Y$.

**Lemma 2.4.8.** *Let $f \colon X \to Y$ be a map of finite simplicial sets. Then the reduced cylinder projection $pr \colon M(f) \to Y$ is a simple map.*

*Proof.* We prove that $pr$ is a simplicial homotopy equivalence over the target, with section the back inclusion $Y \to M(f)$. For each object $(\varphi \colon [n] \to [m], x, y)$ of $\mathrm{simp}(f)$ the composite map $M(N\varphi) \to \Delta^m \to M(N\varphi)$ is induced by the order-preserving function $\psi \colon [n] \sqcup_\varphi [m] \to [n] \sqcup_\varphi [m]$ that maps $[n]$ to $[m]$ by $\varphi$ and is the identity on $[m]$. Then $x \leq \psi(x)$ for all $x \in [n] \sqcup [m]$ in the forward case (resp. $x \geq \psi(x)$ in the backward case), so there is a one-step simplicial homotopy over $\Delta^m$ between the identity on $M(N\varphi)$ and $N\psi$. These maps and the simplicial homotopy are natural in the object $(\varphi, x, y)$, so we may pass to their colimit over $\mathrm{simp}(f)$ to obtain the required simplicial homotopy over $Y$ between the identity on $M(f)$ and the composite map $M(f) \to Y \to M(f)$. $\square$

**Definition 2.4.9.** A map $f \colon X \to Y$ of finite simplicial sets has **simple cylinder reduction** if the (backward) cylinder reduction map $T(f) \to M(f)$ is simple.

**Lemma 2.4.10.** *If the vertical maps in the commutative square*

$$
\begin{array}{ccc}
X & \xrightarrow{\ f\ } & Y \\
\downarrow & & \downarrow \\
X' & \xrightarrow{\ f'\ } & Y'
\end{array}
$$

*are simple, then the induced map $T(f) \to T(f')$ is simple. If furthermore $f$ and $f'$ have simple cylinder reduction, then the induced map $M(f) \to M(f')$ is also simple.*

*Proof.* The case of $T(f) \to T(f')$ is immediate by the gluing lemma for simple maps. With the added hypothesis, it implies the case of $M(f) \to M(f')$ by the composition and right cancellation properties of simple maps, applied to the maps in the commutative square

$$
\begin{array}{ccc}
T(f) & \longrightarrow & T(f') \\
\downarrow & & \downarrow \\
M(f) & \longrightarrow & M(f') .
\end{array}
$$

□

Recall $\mathrm{cone}(X)$ and $\mathrm{simp}_\eta(X)$ from Definition 2.2.13.

**Lemma 2.4.11.** *Let $f\colon X \to Y$ be a map of simplicial sets.*
*(a) There is a natural isomorphism $M(X \to *) \cong \mathrm{cone}(X)$ (in the forward case).*
*(b) The natural map*

$$
(c, pr)\colon M(X \to Y) \to \mathrm{cone}(X) \times Y
$$

*is a cofibration, where $c\colon M(X \to Y) \to \mathrm{cone}(X)$ is the natural map induced by $id_X$ and $Y \to *$.*
*(c) If $X$ and $Y$ are both finite, or both non-singular, then $M(f)$ has the same property.*
*Similar results hold in the backward case, replacing $\mathrm{cone}(X)$ by $\mathrm{cone}^{op}(X) = \mathrm{cone}(X^{op})^{op}$.*

*Proof.* (a) For $f\colon X \to *$, there is a cofinal embedding $j\colon \mathrm{simp}_\eta(X) \to \mathrm{simp}(f)$ that takes the object $([n], x)$ with $x \in X_n$ to the object $(\epsilon\colon [n] \to [0], x, *)$, i.e., the commutative diagram

$$
\begin{array}{ccc}
\Delta^n & \xrightarrow{\ \bar{x}\ } & X \\
{\scriptstyle \epsilon}\downarrow & & \downarrow{\scriptstyle f} \\
\Delta^0 & \longrightarrow & * .
\end{array}
$$

(The restriction of $j$ to $\text{simp}(X) \subset \text{simp}_\eta(X)$ is not cofinal, unless $X$ is path-connected.) The composite of $j$ and the functor $(\varphi, x, y) \mapsto M(N\varphi)$ from the definition of $M(f)$ is naturally isomorphic to the functor $([n], x) \mapsto \text{cone}(\Delta^n)$ from the definition of $\text{cone}(X)$, via the obvious isomorphism $[n] \sqcup_\epsilon [0] \cong [n] \cup \{v\}$. By cofinality [Ma71, IX.3.1], the induced map

$$\text{cone}(X) = \underset{\text{simp}_\eta(X)}{\text{colim}} \left( ([n], x) \mapsto \text{cone}(\Delta^n) \right)$$

$$\xrightarrow{\cong} \underset{\text{simp}(f)}{\text{colim}} \left( (\varphi, x, y) \mapsto M(N\varphi) \right) = M(f)$$

is an isomorphism. The backward case follows by passage to opposites.

(b) Each $q$-simplex $z$ in $M(f)$ is represented by some $q$-simplex $\gamma$ in

$$M(N\varphi) = N([n] \sqcup_\varphi [m]),$$

i.e., an order-preserving function $\gamma \colon [q] \to [n] \sqcup_\varphi [m]$, at some object $(\varphi, x, y)$ of $\text{simp}(f)$. Let $\beta \colon [q] \to [m]$ be the composite of $\gamma$ with $\varphi \vee id \colon [n] \sqcup_\varphi [m] \to [m]$, and let $\alpha \colon [p] \to [n]$ be the pullback of $\gamma$ along $in_1 \colon [n] \to [n] \sqcup_\varphi [m]$. We obtain a commutative diagram

$$
\begin{array}{ccccc}
\Delta^p & \xrightarrow{\alpha} & \Delta^n & \xrightarrow{\bar{x}} & X \\
{\scriptstyle \mu}\downarrow & & {\scriptstyle \varphi}\downarrow & & \downarrow{\scriptstyle f} \\
\Delta^q & \xrightarrow{\beta} & \Delta^m & \xrightarrow{\bar{y}} & Y,
\end{array}
$$

where $-1 \le p \le q$ and $\mu \colon [p] \subset [q]$ is a front face (in the forward case). The left hand square defines a morphism $(\alpha, \beta)$ in $\text{simp}(f)$, so $z$ in $M(f)$ is also represented by the $q$-simplex $\chi$ in $M(N\mu) = N([p] \sqcup_\mu [q])$, given by a canonical order-preserving function $\chi \colon [q] \to [p] \sqcup_\mu [q]$, at the object $(\mu, \alpha^*(x), \beta^*(y))$ of $\text{simp}(f)$. Here $\chi$ maps $j \in [q]$ to $in_1(j)$ for $0 \le j \le p$ and to $in_2(j)$ for $p < j \le q$.

This canonical representative for $z$ is in fact determined by the image of $z$ in $\text{cone}(X) \times Y$. The image $c(z)$ of $z$ in $M(X \to *) \cong \text{cone}(X)$ forgets the maps to $Y$, but remembers $\alpha^*(x)$ and $\mu$. The image $pr(z)$ of $z$ in $Y$ equals $\beta^*(y)$. Taken together, these certainly determine the object $(\mu, \alpha^*(x), \beta^*(y))$. Thus $(c, pr)$ is injective in each simplicial degree $q$.

The backward case is quite similar, replacing front faces by back faces $\mu = (\delta^0)^{q-p} \colon [p] \to [q]$.

(c) If $X$ and $Y$ are both finite, or both non-singular, then $\text{cone}(X)$, $\text{cone}(X) \times Y$ and $M(f)$ all have the same property, by Lemma 2.2.16 and case (b). $\square$

In the remainder of this section we study the reduced mapping cylinder for simplicial maps of the form $N\varphi \colon NC \to ND$, associated to an order-preserving function $\varphi \colon C \to D$ of partially ordered sets. One source of such examples is the function $\varphi = f^\# \colon X^\# \to Y^\#$ associated to a map $f \colon X \to Y$ of simplicial sets, with $N\varphi = B(f)\colon B(X) \to B(Y)$. When $X$ and $Y$ are non-singular, this equals the subdivided map $N\varphi = Sd(f)\colon Sd(X) \to Sd(Y)$. For another kind of example, see Lemma 2.7.4.

Recall from Definition 2.4.3 that $P(\varphi) = C \sqcup_\varphi D$, with either the forward $(c < \varphi(c))$ or backward $(c > \varphi(c))$ partial ordering, according to the context.

**Lemma 2.4.12.** *Let $\varphi\colon C \to D$ be an order-preserving function of partially ordered sets. Then*

$$M(N\varphi) \cong NP(\varphi) = N(C \sqcup_\varphi D),$$

*where $N\varphi\colon NC \to ND$ is the induced map of nerves.*

*In particular, $M(B(f)) \cong N(X^\# \sqcup_{f\#} Y^\#)$ for each map $f\colon X \to Y$ of simplicial sets, and $M(Sd(f)) \cong N(X^\# \sqcup_{f\#} Y^\#)$ for each map $f\colon X \to Y$ of non-singular simplicial sets.*

*Proof.* By the Yoneda lemma, the objects of $\mathrm{simp}(N\varphi)$ correspond bijectively to commutative diagrams

$$
\begin{array}{ccc}
[n] & \xrightarrow{\ x\ } & C \\
{\scriptstyle \psi}\downarrow & & \downarrow{\scriptstyle \varphi} \\
[m] & \xrightarrow{\ y\ } & D
\end{array}
$$

of partially ordered sets and order-preserving functions, which we briefly denote by $(\psi, x, y)$. The colimit of the induced maps $NP(\psi) \to NP(\varphi)$ is the natural map $M(N\varphi) \to NP(\varphi)$ that we assert is an isomorphism.

A $q$-simplex of $NP(\varphi)$ is given by an order-preserving function $w\colon [q] \to C \sqcup_\varphi D$. Let $y\colon [q] \to D$ be its composite with $\varphi \vee id\colon C \sqcup_\varphi D \to D$, and let $x\colon [p] \to C$ be the pullback of $w$ along $in_1\colon C \to C \sqcup_\varphi D$. Then the canonical map $\chi\colon [q] \to [p] \sqcup_\mu [q]$, where $\mu\colon [p] \subset [q]$ is the front face (in the forward case), is a $q$-simplex of $NP(\mu)$ at the object $(\mu, x, y)$ of $\mathrm{simp}(N\varphi)$. The image of $\chi$ in the colimit $M(N\varphi)$ maps to $w$ in $NP(\varphi)$, proving surjectivity.

To prove injectivity, we argue that a $q$-simplex $z$ of $M(N\varphi)$ is determined by its image $w$ in $NP(\varphi)$. Let $\gamma\colon [q] \to [n] \sqcup_\psi [m]$ be a $q$-simplex in $NP(\psi)$, at some object $(\psi, x, y)$ of $\mathrm{simp}(N\varphi)$, that represents $z$ in the colimit. Define $\alpha$, $\beta$ and $\mu\colon [p] \to [q]$ as in the proof of Lemma 2.4.11(b). Then $\chi\colon [q] \to P(\mu)$ is a $q$-simplex in $NP(\mu)$, at the object $(\mu, \alpha^*(x), \beta^*(y))$ of $\mathrm{simp}(N\varphi)$, that also represents $z$. The image simplex $w$ is then equal to the composite map

$$[q] \xrightarrow{\ \chi\ } P(\mu) \xrightarrow{\ \alpha \sqcup \beta\ } P(\psi) \xrightarrow{\ x \sqcup y\ } P(\varphi),$$

which determines $\beta^*(y)\colon [q] \to D$ by composition along $pr\colon P(\varphi) \to D$, and $\mu\colon [p] \to [q]$ and $\alpha^*(x)\colon [p] \to C$ by pullback along $in_1\colon C \to P(\varphi)$. Thus the canonical representative $\chi$ at $(\mu, \alpha^*(x), \beta^*(y))$ of $z$ is, indeed, determined by $w$.

The backward case is similar, replacing front faces $\mu$ by back faces. $\square$

We can also iterate the mapping cylinder construction.

**Definition 2.4.13.** For a chain of $r$ composable maps
$$X_0 \xrightarrow{f_1} \dots \xrightarrow{f_r} X_r$$
of simplicial sets, the **forward $r$-fold iterated** reduced mapping cylinder, denoted $M(f_1, \dots, f_r)$, is recursively defined as the forward reduced mapping cylinder $M(f_r \circ pr)$ of the composite map
$$M(f_1, \dots, f_{r-1}) \xrightarrow{pr} X_{r-1} \xrightarrow{f_r} X_r .$$
The iterated cylinder projection $pr \colon M(f_1, \dots, f_r) \to X_r$ then equals the reduced cylinder projection from $M(f_r \circ pr)$.

Dually, for a chain of $r$ composable maps
$$X_r \xrightarrow{f_r} \dots \xrightarrow{f_1} X_0$$
the **backward $r$-fold iterated** reduced mapping cylinder $M(f_1, \dots, f_r)$ is defined as the backward reduced mapping cylinder of the composite map
$$M(f_2, \dots, f_r) \xrightarrow{pr} X_1 \xrightarrow{f_1} X_0 .$$
The iterated projection $pr \colon M(f_1, \dots, f_r) \to X_0$ equals the reduced cylinder projection from $M(f_1 \circ pr)$.

There is a cylinder coordinate projection $\pi \colon M(f_1, \dots, f_r) \to \Delta^r$, induced by naturality with respect to the terminal map of chains from $X_0 \to \dots \to X_r$ to $* \to \dots \to *$, in the forward case, and similarly in the backward case. The fiber $\pi^{-1}(i)$ of this map is $X_i$, for each $i \in [r]$ viewed as a vertex of $\Delta^r$. The lemmas above imply that if each $X_i$ is finite (resp. non-singular) then $M(f_1, \dots, f_r)$ is also finite (resp. non-singular).

One can also give a direct definition of the $r$-fold iterated reduced mapping cylinder $M(f_1, \dots, f_r)$, as the colimit of a functor to simplicial sets. The relevant functor takes an object in the category $\mathrm{simp}(f_1, \dots, f_r)$, which is a commutative diagram of the form

$$
\begin{array}{ccccc}
\Delta^{n_0} & \xrightarrow{\varphi_1} & \dots & \xrightarrow{\varphi_r} & \Delta^{n_r} \\
\bar{x}_0 \downarrow & & & & \downarrow \bar{x}_r \\
X_0 & \xrightarrow{f_1} & \dots & \xrightarrow{f_r} & X_r ,
\end{array}
$$

to the nerve of the partially ordered set
$$P(\varphi_1, \dots, \varphi_r) = [n_0] \sqcup_{\varphi_1} \dots \sqcup_{\varphi_r} [n_r]$$
(which comes in forward and backward versions). In the forward case its $q$-simplices can be explicitly described as the commutative diagrams

$$
\begin{array}{ccccccc}
\Delta^{q_0} & \xrightarrow{\mu_1} & \dots & \xrightarrow{\mu_{r-1}} & \Delta^{q_{r-1}} & \xrightarrow{\mu_r} & \Delta^q \\
\bar{x}_0 \downarrow & & & & \downarrow \bar{x}_{r-1} & & \downarrow \bar{x}_r \\
X_0 & \xrightarrow{f_1} & \dots & \xrightarrow{f_{r-1}} & X_{r-1} & \xrightarrow{f_r} & X_r
\end{array}
$$

in which the maps $\mu_1, \dots, \mu_r$ are front faces. In the backward case, these maps must be back faces.

**Lemma 2.4.14.** *Let*

$$C_0 \xrightarrow{\varphi_1} \ldots \xrightarrow{\varphi_r} C_r$$

*be a chain of $r$ order-preserving functions of partially ordered sets. Then*

$$M(N\varphi_1, \ldots, N\varphi_r) \cong NP(\varphi_1, \ldots, \varphi_r) = N(C_0 \sqcup_{\varphi_1} \cdots \sqcup_{\varphi_r} C_r),$$

*and similarly in the backward case.*

*Proof.* By induction on $r$ and Lemma 2.4.12 we can compute

$$\begin{aligned} M(N\varphi_1, \ldots, N\varphi_r) &= M(N\varphi_r \circ pr) = M(N\psi) \\ &\cong N(P(\varphi_1, \ldots, \varphi_{r-1}) \sqcup_\psi C_r) = NP(\varphi_1, \ldots, \varphi_r). \end{aligned}$$

Here $\psi \colon C_0 \sqcup_{\varphi_1} \cdots \sqcup_{\varphi_r} C_{r-1} \to C_r$ folds together the composite maps $\varphi_r \circ \cdots \circ \varphi_i \colon C_{i-1} \to C_r$ for $1 \leq i \leq r$. $\square$

To conclude this section, we shall address the question of when the cylinder reduction map $T(f) \to M(f)$ is simple. The example $f = \delta^0 \colon \Delta^0 \to \Delta^1$ shows that $T(f) \to M(f)$ (in the backward versions) need not be surjective, and the example $f = \sigma^0|\partial \colon \partial\Delta^2 \to \Delta^1$ shows that $T(f) \to M(f)$ (again, in the backward versions) can be surjective but not simple.

The applications we have in mind will always concern the backward versions of these mapping cylinders. Therefore, from here on and in the remainder of the book, we will always mean the **backward** reduced mapping cylinder when we write $M(f)$.

**Definition 2.4.15.** A subset $A \subset C$ of a partially ordered set will be called a **left ideal** if it is closed under passage to predecessors, i.e., whenever $a < b$ in $C$ and $b \in A$ then $a \in A$. For each element $v \in C$ let $C/v = \{c \in C \mid c \leq v\}$ be the left ideal generated by $v$. An order-preserving function $\varphi \colon C \to D$ restricts to another such function $C/v \to D/\varphi(v)$. When $v$ is maximal in $C$, $N(C/v) = St(v, NC)$ is the star neighborhood of $v$ in the simplicial complex $NC$.

**Proposition 2.4.16.** *Let $\varphi \colon C \to D$ be an order-preserving function of finite partially ordered sets. Then the following conditions are equivalent:*

*(a) For each element $v \in C$ the induced map $N(C/v) \to N(D/\varphi(v))$ of nerves of left ideals is simple;*

*(b) The map $N\varphi \colon NC \to ND$ has simple backward cylinder reduction.*

*Proof.* We prove that (a) implies (b) by induction over the left ideals $A \subset C$. The inductive statement is that $N(\varphi|A) \colon NA \to ND$ has simple (backward, as always) cylinder reduction. Note that $A/v = C/v$ for each left ideal $A \subset C$ and $v \in A$, so the hypothesis (a) for $\varphi$ implies the corresponding hypothesis for $\varphi|A$. To start the induction, with $A = \emptyset$, note that $T(\emptyset \to ND) \cong M(\emptyset \to ND) \cong ND$.

For the inductive step, let $A \subset C$ be a non-empty left ideal, and assume that $N(\varphi|A')$ has simple cylinder reduction for each left ideal $A' \subset C$ properly contained in $A$.

Choose a maximal element $v \in A$ and let $A' = A \setminus \{v\}$. For brevity let $K = A/v$ and $K' = K \cap A'$, so that $NK = St(v, NA)$ is the star neighborhood of $v$ in $NA$ and $NK' = Lk(v, NA)$ is its link, see Definition 3.2.11. Then $NA = NK \cup_{NK'} NA'$ and $NK = \mathrm{cone}(NK')$ with vertex at $v$. Let $L = D/\varphi(v)$. Then there are two pushout squares

$$(2.4.17) \quad
\begin{array}{ccccc}
T(NK' \to NK) & \longrightarrow & T(NK' \to NL) & \longrightarrow & T(NA' \to ND) \\
\downarrow & & \downarrow & & \downarrow \\
T(NK \to NK) & \longrightarrow & T(NK \to NL) & \longrightarrow & T(NA \to ND)
\end{array}$$

where $T(NK \to NK) = N(K \sqcup K)$ is the cone with vertex $v$ on $T(NK' \to NK) = N(K' \sqcup K)$ (because we are dealing with the backward mapping cylinders). In particular, the vertical maps are cofibrations.

The vertex $v$ is also maximal in the backward partially ordered set $P(\varphi|A) = A \sqcup D$, with left ideal $P(\varphi|A)/v = K \sqcup L$, so there is one more pushout square

$$(2.4.18) \quad
\begin{array}{ccc}
M(NK' \to NL) & \longrightarrow & M(NA' \to ND) \\
\downarrow & & \downarrow \\
M(NK \to NL) & \longrightarrow & M(NA \to ND)
\end{array}$$

where $M(NK \to NL)$ is the cone with vertex $v$ on $M(NK' \to NL)$. Again, the vertical maps are cofibrations.

We shall use the gluing lemma for simple maps for the map from the outer pushout square (2.4.17) to the pushout square (2.4.18). By our inductive hypothesis the cylinder reduction maps $T(NA' \to ND) \to M(NA' \to ND)$ (in the upper right hand corner) and $T(NK' \to ND) \to M(NK' \to ND)$ are simple. The formulas

$$T(NK' \to ND) = T(NK' \to NL) \cup_{NL} ND$$

and

$$M(NK' \to ND) = M(NK' \to NL) \cup_{NL} ND$$

(in the backward case) let us deduce that also the restricted map $T(NK' \to NL) \to M(NK' \to NL)$ is simple. By our hypothesis, condition (a) for $v \in A \subset C$, the map $NK \to NL$ is simple, so by the gluing lemma the map $T(NK' \to NK) \to T(NK' \to NL)$ is simple. Thus the composite map $T(NK' \to NK) \to M(NK' \to NL)$ (in the upper left hand corner) is simple. The map $T(NK \to NK) \to M(NK \to NL)$ (in the lower left hand corner) is the cone on the latter map, with vertex $v$, hence also this third map is simple.

Thus, by the gluing lemma, the map $T(NA \to ND) \to M(NA \to ND)$ (in the lower right hand corner) is a simple map.

This proves that $N(\varphi|A)$ has simple cylinder reduction. By induction, we are done with the proof that (a) implies (b).

For the converse implication (which we will not use elsewhere in the book), suppose that $T(NC \to ND) \to M(NC \to ND)$ is simple and consider any element $v \in C$, with notation as before. By pullback, the cylinder reduction map $T(NK \to NL) \to M(NK \to NL)$ is also simple. Considering its source as a mapping cone, and its target as a cone with vertex $v$, we may restrict to the subspace (after geometric realization) where the cone coordinate is $1/2$. By the pullback property, we deduce that the map $T(NK' \to NK) \to M(NK' \to NL)$ is simple. At the back end of these cylinders we recover the map $NK \to NL$, which must therefore also be simple. This verifies condition (a). $\square$

**Definition 2.4.19.** We say that a map $f\colon X \to Y$ of finite simplicial sets is **simple onto its image** if $f\colon X \to f(X)$ is a simple map, where $f(X) \subset Y$ is the image of $f$.

**Lemma 2.4.20.** *Let $K$ be the simplicial subset generated by an $n$-simplex $x$ of $X$, so $K$ is the image of $\bar{x}\colon \Delta^n \to X$. Then $B(K)$ is the image of $B(\bar{x})$ in $B(X)$, and $Sd(K)$ is the image of $Sd(\bar{x})$.*

*Proof.* Each simplex $y$ of $K$ has the form $\alpha^*(x)$, for some simplicial operator $\alpha$, so the map $Sd(\bar{y})\colon Sd(\Delta^m) \to Sd(K)$ factors through $Sd(\bar{x})$. As $y$ varies, these maps cover $Sd(K)$, hence $Sd(\bar{x})$ maps onto $Sd(K)$. The Barratt nerve case follows by the surjectivity of $Sd(K) \to B(K)$, see Lemma 2.2.10. $\square$

**Lemma 2.4.21.** *Let $f\colon X \to Y$ be a map of finite simplicial sets. Suppose that for each non-degenerate simplex $x \in X^{\#}$ the map $B(\bar{x})\colon B(\Delta^n) \to B(X)$ is simple onto its image, and likewise for each non-degenerate simplex of $Y$. Then $B(f)\colon B(X) \to B(Y)$ has simple backward cylinder reduction.*

For example, the lemma shows that $Sd(f)\colon Sd(X) \to Sd(Y)$ has simple cylinder reduction whenever $X$ and $Y$ are non-singular simplicial sets. See also Lemma 2.5.6 and Corollary 2.5.7.

*Proof.* Let $C = X^{\#}$, $D = Y^{\#}$ and $\varphi = f^{\#}$. For each non-degenerate simplex $v = x \in X^{\#}$ the left ideal $C/v = X^{\#}/x$ is the partially ordered set of non-degenerate faces of $x$, i.e., the non-degenerate simplices $K^{\#}$ of the simplicial subset $K$ of $X$ generated by $x$, so $N(C/v) = B(K)$ equals the image of $B(\bar{x})$ in $B(X)$.

The image $\varphi(v) = y \in Y^{\#}$ satisfies $f(x) = \rho^*(y)$, with $\rho\colon [n] \to [m]$ a degeneracy operator, and $N(D/\varphi(v)) = B(L)$ where $L$ is the simplicial subset of $Y$ generated by $y$. Then in the commutative square

$$
\begin{array}{ccc}
B(\Delta^n) & \xrightarrow{B(\bar{x})} & B(K) \\
{\scriptstyle B(\rho)}\downarrow & & \downarrow{\scriptstyle B(f)} \\
B(\Delta^m) & \xrightarrow{B(\bar{y})} & B(L)
\end{array}
$$

the left hand vertical map $B(\rho)$ is a simplicial homotopy equivalence over the target (with simplicial section $B(\hat\rho)$), thus simple. The upper and lower horizontal maps $B(\bar x)$ and $B(\bar y)$ are simple onto their images by hypothesis, so the right hand vertical map $N\varphi = B(f) \colon B(K) \to B(L)$ is simple by the composition and right cancellation properties of simple maps.

It therefore follows from Proposition 2.4.16 that the natural map $T(B(f)) \to M(B(f))$ is simple. $\square$

### 2.5. Making simplicial sets non-singular

In this section, we show that finite simplicial sets can be replaced by non-singular ones, up to simple maps. We first offer an inductive, non-functorial construction, in Proposition 2.5.1. Thereafter, in Theorem 2.5.2, we provide a functorial construction $I(X) = B(Sd(X))$, called the **improvement** functor, with a natural simple map to $X$. The verification that $I$ preserves simple maps relies on the more complicated part of Proposition 2.3.3, for not necessarily non-singular simplicial sets.

**Proposition 2.5.1.** *(a) Let $Y$ be a finite simplicial set. There exists a finite non-singular simplicial set $Z$, and a simple map $f \colon Z \to Y$.*

*(b) More generally, let $X$ and $Y$ be finite simplicial sets. Let $X$ be non-singular, and let $y \colon X \to Y$ be a cofibration. Then there exists a cofibration $z \colon X \to Z$, to a finite non-singular simplicial set $Z$, and a simple map $f \colon Z \to Y$ with $fz = y$. If $y$ is a weak homotopy equivalence, then so is $z$.*

*Proof.* (a) By induction, we may assume that the lemma is proved for $Y$, so that we have a simple map $f \colon Z \to Y$ with $Z$ finite and non-singular, and we need to prove the lemma for $Y' = \Delta^n \cup_{\partial\Delta^n} Y$. Let $W = \partial\Delta^n \times_Y Z$ be the pullback. It is finite and non-singular, because it is contained in the product $\partial\Delta^n \times Z$. Let $g$ and $h$ be the structure maps from $W$ to $\partial\Delta^n$ and $Z$, respectively. Then there is a commutative diagram

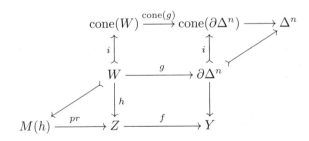

where the horizontal maps are simple, by Proposition 2.1.3(c), the proof of Proposition 2.2.18 and Lemma 2.4.8. By the gluing lemma, the induced map $f' \colon Z' \to Y'$ of pushouts is simple, where

$$Z' = \mathrm{cone}(W) \cup_W M(h).$$

Here $Z'$ is a finite non-singular simplicial set, because $i\colon W \to \mathrm{cone}(W)$ and the front inclusion $W \to M(h)$ are both cofibrations.

(b) In the relative case, we proceed by induction as before and assume given a finite cofibration $z\colon X \to Z$ with $fz = y$. The required cofibration $z'\colon X \to Z'$ is then defined as the composite of the cofibration $z\colon X \to Z$, the back inclusion $Z \to M(h)$ and the inclusion $M(h) \to Z'$.

Finally, if $y\colon X \to Y$ is a weak homotopy equivalence, and $f\colon Z \to Y$ is simple, then any $z\colon X \to Z$ with $fz = y$ is a weak homotopy equivalence, by the 2-out-of-3 property. $\square$

**Theorem 2.5.2.** *There is an improvement functor $I\colon \mathcal{C} \to \mathcal{D}$, which takes each finite simplicial set $X$ to the finite non-singular simplicial set*

$$I(X) = B(Sd(X)) = B(Sd^{op}(X)),$$

*and a natural simple map $s_X\colon I(X) \to X$. The latter map is characterized by the commutative square*

$$
\begin{array}{ccc}
Sd(Sd^{op}(X)) & \xrightarrow{\;d_{Sd^{op}(X)}\;} & Sd^{op}(X) \\
{\scriptstyle b_{Sd^{op}(X)}}\big\downarrow & & \big\downarrow{\scriptstyle d_X^{op}} \\
B(Sd^{op}(X)) & \xrightarrow{\;\;s_X\;\;} & X
\end{array}
$$

*of simple maps. The functor $I$ preserves simple maps, weak homotopy equivalences and cofibrations.*

The proof will be given at the end of this section. The non-degenerate simplices of $Sd(X)$ and $Sd^{op}(X)$ are the same, hence the equality $B(Sd(X)) = B(Sd^{op}(X))$. Proposition 2.5.1 would follow immediately from Theorem 2.5.2, taking $Z = I(Y)$ with its (natural) simple map to $Y$ in case (a), and

$$Z = I(Y) \cup_{I(X)} M(I(X) \to X)$$

in case (b), but the inductive proof we gave is more direct.

Not every CW-complex is triangulable, i.e., homeomorphic to the underlying polyhedron of some simplicial complex. See [FP90, §3.4] for an example due to W. Metzler [Me67] of a non-triangulable finite CW-complex. A sufficient condition for a CW-complex $X$ to be triangulable is that it is **regular**, i.e., that the closure of each open $n$-cell in $X$ is an $n$-ball and its boundary in the closure is an $(n-1)$-sphere. If desired, see §1.1, §1.2 and Theorem 3.4.1 of [FP90] for more details.

There is also a notion of **regularity** for simplicial sets [FP90, p. 208], and the geometric realization of a regular simplicial set is a regular CW-complex and thus triangulable [FP90, Prop. 4.6.11]. The normal subdivision of any simplicial set is a regular simplicial set [FP90, Prop. 4.6.10].

We shall instead make use of the simplicial opposite of the normal subdivision, the op-normal subdivision, which produces simplicial sets with the property opposite to regularity. We shall call this property op-regularity, so that $X$ is op-regular if and only if the opposite of $X$ is regular.

**Definition 2.5.3.** Let $x$ be a non-degenerate $n$-simplex in a simplicial set $X$, with $n \geq 1$, and consider the simplicial subsets $Y \subset X$ and $Z \subset Y$, generated by $x$ and its 0-th face $z = d_0(x)$, respectively. The $n$-simplex $x$ is **op-regular** if $Y$ is obtained from $Z$ by attaching $\Delta^n$ along its 0-th face $\delta^0 \colon \Delta^{n-1} \to \Delta^n$ via the representing map $\bar{z} \colon \Delta^{n-1} \to Z$. In other words, $x$ is op-regular if the diagram

$$
\begin{array}{ccc}
\Delta^{n-1} & \xrightarrow{\;\delta^0\;} & \Delta^n \\
\bar{z} \downarrow & & \downarrow \bar{x} \\
Z & \longrightarrow & Y
\end{array}
$$

is a pushout square. (We can then write $Y = \Delta^n \cup_{\Delta^{n-1}} Z$, but this hides the assumption that $\Delta^{n-1}$ is included as the 0-th face.) By definition, every 0-simplex is op-regular. A simplicial set $X$ is **op-regular** if every non-degenerate simplex of $X$ is op-regular.

Any non-singular simplicial set is op-regular, but not conversely, because the non-degenerate simplices of an op-regular simplicial set may fail to be embedded along their 0-th face.

*Remark 2.5.4.* Simplicial subsets and arbitrary products of op-regular simplicial sets are again op-regular. It follows that to each simplicial set $X$ there is an initial map $X \to R^{op}X$ onto an op-regular simplicial set, which we might call its **op-regularization**. Compare Remark 2.2.12.

To see the claim concerning products, let $(x, y)$ be a non-degenerate simplex of $X \times Y$, with $X$ and $Y$ op-regular. Then $x$ generates a simplicial subset $\Delta^n \cup_{\Delta^{n-1}} Z$ of $X$ and $y$ generates a simplicial subset $\Delta^m \cup_{\Delta^{m-1}} W$ of $Y$, with conventions as previously stated. Then $(x, y)$ is the image under

$$
\Delta^n \times \Delta^m \to (\Delta^n \cup_{\Delta^{n-1}} Z) \times (\Delta^m \cup_{\Delta^{m-1}} W)
$$

of a non-degenerate simplex in the source. Only its 0-th face lies in $\Delta^n \times \delta^0(\Delta^{m-1}) \cup \delta^0(\Delta^{n-1}) \times \Delta^m$, so $(x, y)$ is op-regular. The argument for general products is similar.

**Lemma 2.5.5.** *For every simplicial set $X$, the op-normal subdivision $Sd^{op}(X)$ is an op-regular simplicial set.*

*Proof.* This is the opposite of [FP90, Prop. 4.6.10]. We may inductively assume that $X = \Delta^m \cup_{\partial \Delta^m} X'$ for some map $\bar{y} \colon \Delta^m \to X$, and that $Sd^{op}(X')$ is op-regular. Each non-degenerate $n$-simplex $x$ in $Sd^{op}(X)$ that is not in $Sd^{op}(X')$ is in the image from $Sd^{op}(\Delta^m)$, but not in the image from $Sd^{op}(\partial \Delta^m)$. Hence $x$ is the image under $Sd^{op}(\bar{y})$ of a unique non-degenerate $n$-simplex $(\iota_m = \mu_0 > \cdots > \mu_n)$ in $Sd^{op}(\Delta^m)$, with 0-th vertex at the "barycenter" $(\iota_m)$ and 0-th face in $Sd^{op}(\partial \Delta^m)$. Then the simplicial subset $Y \subset Sd^{op}(X)$ generated by $x$ is the pushout of the diagram

$$
\Delta^n \xleftarrow{\;\delta^0\;} \Delta^{n-1} \to Z
$$

where $Z \subset Y \cap Sd^{op}(X')$ is the image of the face $(\mu_1 > \cdots > \mu_n)$. Hence $x$ is op-regular. □

The lemma just proved will provide our source of op-regular simplicial sets.

**Lemma 2.5.6.** *Let $X$ be an op-regular simplicial set and $x$ a simplex of $X$. Then the maps $\bar{x}\colon \Delta^n \to X$, $B(\bar{x})\colon B(\Delta^n) \to B(X)$ and $Sd(\bar{x})\colon Sd(\Delta^n) \to Sd(X)$ are simple onto their respective images.*

*Proof.* In each case we may assume that $x$ is non-degenerate, for if $x = \rho^*(x^\#)$ with $\rho\colon [n] \to [m]$ a degeneracy operator, then each of $\rho$, $B(\rho)$ and $Sd(\rho)$ is a simplicial homotopy equivalence over the target, with section (induced by) $\hat{\rho}$, and therefore a simple map. The conclusion for $x$ then follows from the one for $x^\#$ by the composition property.

Suppose, then, that $x$ is a non-degenerate $n$-simplex in the op-regular simplicial set $X$. The image $Y$ of $\bar{x}$ is a pushout $\Delta^n \cup_{\Delta^{n-1}} Z$. By induction on $n$, we may assume that $\bar{z}\colon \Delta^{n-1} \to Z$ is simple. Hence $\bar{x}$ is simple, by the gluing lemma for simple maps.

The same argument works for the map $Sd(\bar{x})$ and its image.

In the Barratt nerve case, we may also assume by induction that

$$B(\bar{z})\colon B(\Delta^{n-1}) \to B(Z)$$

is a simple map. Consider the decomposition

$$B(\Delta^n) \cong \mathrm{cone}^{op}(B(\Delta^{n-1})) \cup_{B(\Delta^{n-1})} B(\Delta^{n-1}) \times \Delta^1 \,,$$

where the cone point corresponds to the vertex $0$ in $\Delta^n$. There is a similar decomposition

$$B(Y) \cong \mathrm{cone}^{op}(B(\Delta^{n-1})) \cup_{B(\Delta^{n-1})} M(B(\bar{z})) \,,$$

coming from a poset decomposition $Y^\# \cong \{0\} \sqcup (\Delta^{n-1})^\# \sqcup Z^\#$. The map $B(\bar{x})$ is the identity on the cone, and maps the cylinder by the composite map

$$B(\Delta^{n-1}) \times \Delta^1 \to T(B(\bar{z})) \to M(B(\bar{z})) \,.$$

The first map is a pushout along $B(\bar{z})$, which is simple by our inductive hypothesis. The second map is the natural cylinder reduction map of backward mapping cylinders for the map $B(\bar{z})$, which is simple by Lemma 2.4.21. The hypothesis of that lemma is satisfied in this case, again by the inductive hypothesis, because each non-degenerate simplex of $Z$ is of dimension $< n$. □

**Corollary 2.5.7.** *If $f\colon X \to Y$ is a map of op-regular simplicial sets, then $g = B(f)\colon B(X) \to B(Y)$ has simple backward cylinder reduction.*

*Proof.* This is immediate from Lemmas 2.4.21 and 2.5.6. □

We now turn to a closer comparison of the normal subdivision and Barratt nerve of (op-)regular simplicial sets. Some assumption such as op-regularity is required for the following factorization. For example, there is no such factorization for $X = \Delta^n/\partial\Delta^n$ with $n \geq 1$.

**Proposition 2.5.8.** *Let $X$ be a finite op-regular simplicial set. The last vertex map $d_X \colon Sd(X) \to X$ factors uniquely as the composite of the canonical map $b_X \colon Sd(X) \to B(X)$ and a map $c_X \colon B(X) \to X$.*

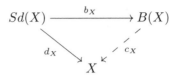

*All three maps in this diagram are simple.*

*Proof.* For the last assertion it suffices to prove that $b_X$ is simple, because the last vertex map $d_X$ is simple by Proposition 2.2.18, so the simplicity of the third map $c_X$ will follow by the right cancellation property of simple maps. Uniqueness of the factorization is also immediate once we know that $b_X$ is simple, hence surjective.

The proof proceeds by induction on the dimension $n$ of the op-regular simplicial set $X$, and a second induction on the number of non-degenerate simplices of $X$. Let $x \in X_n$ be non-degenerate. There is a decomposition

$$X = Y \cup_{Y'} X'$$

where $Y \subset X$ is the simplicial subset generated by $x$, and $Y' = Y \cap X'$ is of dimension $< n$. Then $Sd(X) = Sd(Y) \cup_{Sd(Y')} Sd(X')$ and $B(X) = B(Y) \cup_{B(Y')} B(X')$.

If $Y$ is properly contained in $X$, then by induction the proposition holds for $Y$, $Y'$ and $X'$. The uniqueness of the factorization for $Y'$ ensures that we can define $c_X$ as the union of $c_Y$ and $c_{X'}$ along $c_{Y'}$. The simplicity of the map $b_X$ follows from that of $b_Y$, $b_{Y'}$ and $b_{X'}$ by the gluing lemma for simple maps.

It remains to consider the special case when $X$ is generated by a single $n$-simplex, so $X = Y$. Then $X$ is the pushout of a diagram

$$\Delta^n \xleftarrow{\delta^0} \Delta^{n-1} \xrightarrow{\bar{z}} Z$$

as in the definition of "op-regular," where $Z \subset X$ is op-regular of lower dimension. We consider the decomposition

$$Sd(\Delta^n) = \mathrm{cone}^{op}(Sd(\Delta^{n-1})) \cup_{Sd(\Delta^{n-1})} Sd(\Delta^{n-1}) \times \Delta^1$$

where the cone point corresponds to the vertex $0$ of $\Delta^n$. When restricted to the part $Sd(\Delta^{n-1}) \times \Delta^1$ of $Sd(\Delta^n)$ the last vertex map $d_{\Delta^n}$ factors as the composite

$$Sd(\Delta^{n-1}) \times \Delta^1 \xrightarrow{pr_1} Sd(\Delta^{n-1}) \xrightarrow{d_{\Delta^{n-1}}} \Delta^{n-1} \xrightarrow{\delta^0} \Delta^n \,.$$

From $Sd(X) = Sd(\Delta^n) \cup_{Sd(\Delta^{n-1})} Sd(Z)$ we obtain the corresponding decomposition

(2.5.9) $$Sd(X) = \text{cone}^{op}(Sd(\Delta^{n-1})) \cup_{Sd(\Delta^{n-1})} X''$$

where

$$X'' = Sd(\Delta^{n-1}) \times \Delta^1 \cup_{Sd(\Delta^{n-1})} Sd(Z)$$

is the backward version of the mapping cylinder $T(Sd(\bar{z}))$ of the subdivided map $Sd(\bar{z}) \colon Sd(\Delta^{n-1}) \to Sd(Z)$. It follows (from the factorization above) that on the part $X''$ of $Sd(X)$ the last vertex map $d_X$ factors as the composite

$$X'' = T(Sd(\bar{z})) \xrightarrow{pr} Sd(Z) \xrightarrow{d_Z} Z \to X.$$

Here $pr$ denotes the cylinder projection induced by $pr_1$ and the map $Z \to X$ is the inclusion induced by $\delta^0$.

There is a similar decomposition of $B(X) = N(X^{\#})$, where $X^{\#}$ is the partially ordered set of non-degenerate simplices in $X$. Let $v \in X^{\#}$ be the image under $\bar{x} \colon \Delta^n \to X$ of the vertex 0, i.e., the cone point considered above. Then we can decompose $X^{\#}$ as

$$X^{\#} \cong \{v\} \sqcup (\Delta^{n-1})^{\#} \sqcup Z^{\#},$$

where $(\Delta^{n-1})^{\#}$ corresponds to the non-degenerate simplices of $X$ that properly contain the vertex $v$. The partial ordering on $X^{\#}$ is such that $v < \mu$ for each element $\mu$ of $(\Delta^{n-1})^{\#}$, and $\mu > (\bar{z}^{\#})(\mu)$, where $\bar{z}^{\#} \colon (\Delta^{n-1})^{\#} \to Z^{\#}$. Hence

(2.5.10) $$B(X) = \text{cone}^{op}(B(\Delta^{n-1})) \cup_{B(\Delta^{n-1})} B''$$

where $B'' = NP(\bar{z}^{\#})$ is the nerve of the backward version of the partially ordered set

$$P(\bar{z}^{\#}) = (\Delta^{n-1})^{\#} \sqcup_{\bar{z}^{\#}} Z^{\#}.$$

By Lemma 2.4.12, we can write $B'' = M(B(\bar{z}))$ as the backward version of the reduced mapping cylinder of the Barratt nerve map $B(\bar{z}) \colon B(\Delta^{n-1}) \to B(Z)$.

The canonical map $b_X$ takes $X''$ to $B''$, and comparing (2.5.9) and (2.5.10) we see that the left hand square in the following diagram is a pushout.

(2.5.11)

$$
\begin{array}{ccccccc}
Sd(X) & \longleftarrow & X'' = T(Sd(\bar{z})) & \xrightarrow{pr} & Sd(Z) & \xrightarrow{d_Z} & Z \\
{\scriptstyle b_X}\downarrow & & \downarrow & & {\scriptstyle b_Z}\downarrow\;\;\nearrow{\scriptstyle c_Z} & & \downarrow \\
B(X) & \longleftarrow & B'' = M(B(\bar{z})) & \xrightarrow{pr} & B(Z) & & X
\end{array}
$$

The unlabeled vertical map $X'' \to B''$ is the composite of two simple maps. The first map $X'' = T(Sd(\bar{z})) \to T(B(\bar{z}))$ is the natural map of mapping cylinders induced by the commutative square

$$
\begin{array}{ccc}
Sd(\Delta^{n-1}) & \xrightarrow{Sd(\bar{z})} & Sd(Z) \\
{\scriptstyle =}\downarrow & & \downarrow{\scriptstyle b_Z} \\
B(\Delta^{n-1}) & \xrightarrow{B(\bar{z})} & B(Z),
\end{array}
$$

and is simple by the gluing lemma and the inductive hypothesis that $b_Z$ is simple. The second map $T(B(\bar{z})) \to M(B(\bar{z})) = B''$ is the natural cylinder reduction map of backward mapping cylinders, and is simple by Corollary 2.5.7, because $\Delta^{n-1}$ is non-singular and $Z$ is op-regular.

It follows that the composite map $X'' \to B''$ is simple, so by the gluing lemma for simple maps applied to the pushout square in (2.5.11), it follows that $b_X \colon Sd(X) \to B(X)$ is a simple map. Next, the compatibility of the cylinder projections from the ordinary and the reduced mapping cylinder ensures that the middle square in (2.5.11) commutes. By the inductive hypothesis on $Z$ there is a map $c_Z$ that factors $d_Z$ through $b_Z$. Hence the restriction of the last vertex map $d_X$ to $X''$, which is given by the composite along the upper and right hand edge of (2.5.11), factors through the simple map $X'' \to B''$. By the universal property of a categorical pushout, it follows that $d_X \colon Sd(X) \to X$ itself factors through $b_X \colon Sd(X) \to B(X)$, as claimed.  $\square$

*Proof of Theorem 2.5.2.* The improvement functor $I$ is defined as the composite $B \circ Sd = B \circ Sd^{op}$. It preserves cofibrations, because $B$ and $Sd$ do. It also preserves simple maps. For if $f \colon X \to Y$ is simple, then $Sd^{op}(f) \colon Sd^{op}(X) \to Sd^{op}(Y)$ is a simple map by Proposition 2.3.3 and the remark that $Sd(X)$ and $Sd^{op}(X)$ have naturally homeomorphic geometric realizations. Hence also $Sd(Sd^{op}(f)) \colon Sd(Sd^{op}(X)) \to Sd(Sd^{op}(Y))$ is simple, by Proposition 2.3.3 again. The vertical maps in the commutative square

$$
\begin{array}{ccc}
Sd(Sd^{op}(X)) & \xrightarrow{\ Sd(Sd^{op}(f))\ } & Sd(Sd^{op}(Y)) \\
\Big\downarrow{\scriptstyle b_{Sd^{op}(X)}} & & \Big\downarrow{\scriptstyle b_{Sd^{op}(Y)}} \\
B(Sd^{op}(X)) & \xrightarrow{\quad I(f)\quad} & B(Sd^{op}(Y))
\end{array}
$$

are simple, by Proposition 2.5.8 applied to the finite simplicial sets $Sd^{op}(X)$ and $Sd^{op}(Y)$, respectively, which are op-regular by Lemma 2.5.5. Thus the lower horizontal map $I(f) \colon I(X) \to I(Y)$ is simple, by the composition and right cancellation properties of simple maps.

Finally, there is a natural transformation $s_X \colon I(X) \to X$. It is defined as the composite map

$$s_X \colon I(X) = B \circ Sd^{op}(X) \xrightarrow{\ c_{Sd^{op}(X)}\ } Sd^{op}(X) \xrightarrow{\ d_X^{op}\ } X \,,$$

where the first map $c_{Sd^{op}(X)}$ is provided by the factorization in Proposition 2.5.8 of the last vertex map for the op-regular simplicial set $Sd^{op}(X)$, and the second map $d_X^{op}$ is the first vertex map for $X$, from Definition 2.2.20. Each of these is simple, so also $s_X$ is a simple map. It follows by the 2-out-of-3 property that $I$ also preserves weak homotopy equivalences.  $\square$

## 2.6. THE APPROXIMATE LIFTING PROPERTY

The following study of the ALP will be useful in Section 2.7 when we relate Serre fibrations to simple maps.

**Definition 2.6.1.** Let $f\colon X \to Y$ be a map of topological spaces. We say that $f$ has the **approximate lifting property for polyhedra**, abbreviated ALP, if for each commutative diagram of solid arrows of the following kind

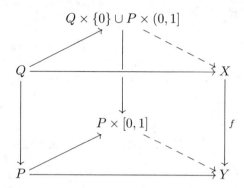

the dashed arrows can always be filled in, keeping the diagram commutative. Here $P$ is a compact polyhedron and $Q$ a compact subpolyhedron, we identify $P$ with $P \times \{0\} \subset P \times [0,1]$ and $Q$ with $Q \times \{0\}$, and $(0,1]$ is the half-open interval containing 1 but not 0. We give $Q \times \{0\} \cup P \times (0,1]$ the quotient topology from $Q \times [0,1] \coprod P \times (0,1]$, which equals the compactly generated topology associated with the subspace topology from $P \times [0,1]$.

We say that $f\colon X \to Y$ has the **relative approximate lifting property** if, moreover, the restriction to $Q \times [0,1]$ of the approximate lifting map $Q \times \{0\} \cup P \times (0,1] \to X$ can be arranged to be constant in the interval coordinate.

In other words, given any map $g\colon P \to Y$, there should exist a homotopy $G\colon P \times [0,1] \to Y$ starting with $g$, such that the remainder $P \times (0,1] \to Y$ of the homotopy admits a lift $P \times (0,1] \to X$ over $f\colon X \to Y$. Furthermore, if we are given a lift $h\colon Q \to X$ over $f$ of the restriction of $g$ to $Q$, then the union $H\colon Q \times \{0\} \cup P \times (0,1] \to X$ of the two lifts is required to be a continuous map. In the relative case we also ask that the lifted homotopy is constant on $Q$, so that $H$ restricted to $Q \times [0,1]$ equals $h \circ pr_1$. This is only seemingly stronger than the absolute ALP, as the following lemma shows.

**Lemma 2.6.2.** *The ALP and relative ALP are equivalent.*

*Proof.* Suppose that $f\colon X \to Y$ has the ALP, let $(P,Q)$ be a pair of compact polyhedra, and let $g\colon P \to X$ and $h\colon Q \to Y$ be maps with $g|Q = fh$. Apply the ALP for the polyhedral pair $(P \times [0,1], Q \times [0,1])$ and the maps $g \circ pr_1$ and $h \circ pr_1$, to get a homotopy $G'\colon P \times [0,1] \times [0,1] \to Y$ of $g \circ pr_1$, with an approximate lift $H'\colon Q \times [0,1] \times \{0\} \cup P \times [0,1] \times (0,1] \to X$ such that $H'(q,t,0) = h(q)$. Let $\epsilon\colon P \times [0,1] \to [0,1]$ be a continuous function with $\epsilon^{-1}(0) = Q \times [0,1] \cup P \times \{0\}$. Then $G\colon P \times [0,1] \to Y$ given by $G(p,t) = G'(p,t,\epsilon(p,t))$ is a homotopy of $g$, with an approximate lift $H\colon Q \times \{0\} \cup P \times (0,1] \to X$ given by $H(p,t) = H'(p,t,\epsilon(p,t))$, whose restriction to $Q \times [0,1]$ equals $h \circ pr_1$, i.e., is constant in the interval coordinate. Thus $f$ satisfies the relative ALP. $\square$

**Definition 2.6.3.** Let $f\colon X \to Y$ be a map of topological spaces. We say that $f$ is a **simple map** if for each open subset $U \subset Y$ the restricted map $f^{-1}(U) \to U$ is a weak homotopy equivalence, i.e., if $f$ is a hereditary weak homotopy equivalence.

By Proposition 2.1.2 a map $f\colon X \to Y$ of finite simplicial sets is simple, in the sense that its geometric realization $|f|\colon |X| \to |Y|$ has contractible point inverses, if and only if $|f|$ is a simple map of topological spaces in the sense just defined. However, a general simple map of topological spaces need not even be surjective, as the example of the inclusion $(0,1] \subset [0,1]$ illustrates, even if the image will be dense.

The approximate lifting property for polyhedra implies the property of being simple, and for maps to metric spaces the two notions are equivalent, as we shall momentarily prove.

The ALP can also be motivated by considering what is required for the cylinder coordinate projection $\pi\colon T(f) = X \times [0,1] \cup_X Y \to [0,1]$ to be a Serre fibration. For details, see the proof of Proposition 2.6.10.

**Lemma 2.6.4.** *If $f\colon X \to Y$ has the ALP, then it is simple.*

*Proof.* Let $U \subset Y$ be an open subset. Given any commutative square of solid arrows

$$
\begin{array}{ccc}
S^{n-1} & \longrightarrow & f^{-1}(U) \\
\downarrow & \nearrow & \downarrow \\
D^n & \longrightarrow & U
\end{array}
$$

we shall show that there is a map $D^n \to f^{-1}(U)$ such that the resulting upper triangle commutes and the lower triangle commutes up to homotopy. This implies that the restricted map $f^{-1}(U) \to U$ is a weak homotopy equivalence, and that $f$ is a simple map.

We apply the ALP for $f\colon X \to Y$ and a polyhedral pair $(P,Q)$ homeomorphic to $(D^n, S^{n-1})$, to obtain the dashed arrows in the following commutative diagram.

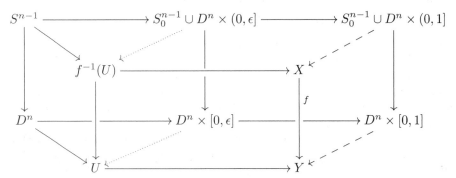

Here we have abbreviated $S^{n-1} \times \{0\}$ to $S_0^{n-1}$ two times in the top row, for typographical reasons. Next we note that these dashed arrows to $X$ and $Y$ will restrict to dotted arrows mapping to $f^{-1}(U)$ and $U$, respectively, provided we make $\epsilon > 0$ sufficiently small. The required map can then be obtained as the composite

$$D^n \to S^{n-1} \times \{0\} \cup D^n \times (0, \epsilon] \to f^{-1}(U),$$

where the first map is chosen to extend the identification $S^{n-1} \cong S^{n-1} \times \{0\}$. $\square$

**Proposition 2.6.5.** *If $f \colon X \to Y$ is simple and $(Y, d)$ is a metric space, then $f$ has the relative ALP.*

*Proof.* Let $P$ be a compact polyhedron and $g \colon P \to Y$ a map. For the purpose of this proof let an $\epsilon$-**lifting** of $g$ be a map $h \colon P \to X$ such that for each point $x \in P$ the distance in $Y$ from $g(x)$ to $f(h(x))$ is less than $\epsilon$. Let $T$ be a triangulation of $P$, i.e., an ordered simplicial complex (or if preferred, a non-singular simplicial set) with a PL homeomorphism $|T| \cong P$. By a $(T, \epsilon)$-**lifting** of $g$ we shall mean a map $h \colon P \to X$ such that for each simplex $S \subset P$ in $T$ the image $f(h(S))$ is contained in the open $\epsilon$-neighborhood in $Y$ of $g(S)$.

Let the **mesh** of $(P, T, g)$ be the number $\tau$ given as the maximum of the distances $d(g(x_1), g(x_2))$, where $x_1$ and $x_2$ are any two points in $P$ that are contained in a single simplex in $T$. So $\tau$ is the maximum of the diameters of the images $g(S) \subset Y$ as $S \subset P$ ranges through the simplices of the triangulation $T$. Then any $(T, \epsilon)$-lifting will be a $(\tau + \epsilon)$-lifting, by the triangle inequality. Conversely, an $\epsilon$-lifting is a $(T, \epsilon)$-lifting for any triangulation $T$.

We claim that for any triangulation $T$ of $P$ and any $\epsilon > 0$ it is possible to find a $(T, \epsilon)$-lifting of $g \colon P \to Y$. Moreover, any given $(T, \epsilon)$-lifting on a subcomplex $P'$ of $P$ (triangulated by a subcomplex $T'$ of $T$) can be extended to one on all of $P$.

By induction it will suffice to prove this claim in the case when $P'$ is all of $P$ except from the interior of a top-dimensional simplex. This case again follows from the special case where $P$ is a simplex and $P'$ is its boundary.

In this case, let $U$ be the open $\epsilon$-neighborhood in $Y$ of the image $g(P)$. The partial $(T, \epsilon)$-lifting on $P'$ may be regarded as a map to $f^{-1}(U)$, in view of the very definition of a $(T, \epsilon)$-lifting. We are assuming that $f$ is a simple map, so that the restricted map $f^{-1}(U) \to U$ is a weak homotopy equivalence. Hence we can find a map $P \to f^{-1}(U)$ extending the given map on $P'$. This gives the desired $(T, \epsilon)$-lifting on $P$, proving the claim.

To prove that $f$ has the relative ALP we shall construct a map $P \times (0, 1] \to X$ by piecing together liftings of the kind just discussed on the parts $P \times \{a_n\}$ and $P \times [a_{n+1}, a_n]$, where $\{a_n\}_{n=1}^{\infty}$ is some sequence of numbers in $(0, 1]$ starting with $a_1 = 1$ and decreasing monotonically to zero, for example $a_n = 1/n$. We must arrange this so that the union maps $Q \times \{0\} \cup P \times (0, 1] \to X$ and $P \times [0, 1] \to Y$ are continuous.

First, for any $n$ we choose a triangulation $T_n$ of $P \times \{a_n\}$ whose mesh, in the above sense, is at most $1/n$. We may certainly arrange that $Q \cong Q \times \{a_n\}$ is triangulated by a subcomplex of $T_n$. Then we use the prior claim to define a map $P \cong P \times \{a_n\} \to X$, extending the given map $Q \to X$ to a $(T_n, 1/n)$-lifting of $g \colon P \to Y$.

Next, we choose a triangulation $T'_n$ of $P \times [a_{n+1}, a_n]$ of mesh $\leq 1/n$. Again we may assume that $Q \times [a_{n+1}, a_n]$ is a subcomplex in this triangulation. The $n$-th and $(n+1)$-th lifting just constructed may then be regarded as a single $(2/n)$-lifting defined on $P \times \{a_{n+1}, a_n\}$, which we may extend to $P \times \{a_{n+1}, a_n\} \cup Q \times [a_{n+1}, a_n]$ by insisting that the lifting on $Q \times [a_{n+1}, a_n]$ must be independent of the interval coordinate. By the prior claim, again, the lifting may be further extended to a $(T'_n, 2/n)$-lifting on all of $P \times [a_{n+1}, a_n]$.

We have thus obtained, for every $n$, a $(3/n)$-lifting on $P \times [a_{n+1}, a_n]$, and these liftings are compatible as $n$ varies. Thus they assemble to a map $P \times (0, 1] \to X$. In view of its construction, this map is compatible with the given maps $Q \to X$ and $P \to Y$ in the way required by the relative ALP. $\square$

**Definition 2.6.6.** Let $p \colon E \to B$ be a map of topological spaces, so $(E, p)$ is a space over $B$. We say that $(E, p)$ is **trivial up to a simple map** if there is a proper and simple map over $B$ from a product fibration $(B \times F, pr_1)$ to $(E, p)$. In other words, there is a space $F$ and a commutative diagram

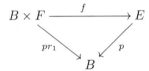

where $f$ is proper and simple.

We say that $(E, p)$ is **locally trivial up to simple maps** if each point in $B$ has a neighborhood over which the restriction of $(E, p)$ is trivial up to a simple map.

**Proposition 2.6.7.** *Let $(E, p)$ be a space over $B$ that is locally trivial up to simple maps and suppose that $E$ is a locally compact metric space. Then $p \colon E \to B$ is a Serre fibration.*

*Proof.* We must show that $p$ has the homotopy lifting property for polyhedra, that is, given any commutative square of solid arrows

(2.6.8)
$$
\begin{array}{ccc}
P & \xrightarrow{\ g\ } & E \\
\downarrow & \nearrow & \downarrow{\scriptstyle p} \\
P \times [0, 1] & \xrightarrow[\ H\ ]{} & B
\end{array}
$$

where $P$ is a compact polyhedron and $P$ is identified with $P \times \{0\}$, the dashed arrow can always be filled in while keeping the diagram commutative. The first step will be to reduce to the case when $(E, p)$ is (globally) trivial up to a simple map.

The base $B$ is covered by open subsets over each of which the restriction of $(E, p)$ is trivial up to a simple map. We can find subdivisions of $P$ and $[0, 1]$ such that each of the blocks $S \times [a, b]$, where $S$ is a simplex of the subdivision of $P$ and $[a, b]$ a simplex of the subdivision of $[0, 1]$, is mapped by the homotopy $H$ into one of these open subsets of $B$. By induction over these blocks it will then suffice to establish that $p$ has the (seemingly stronger, but actually equivalent) relative form of the homotopy lifting property for polyhedra. This relative form asks, in particular, that the dashed arrow can always be filled in the following solid arrow diagram

$$S \times \{a\} \cup \partial S \times [a, b] \longrightarrow E$$

with vertical maps, $H$, $p$, and

$$S \times [a, b] \xrightarrow{H} B .$$

Here $\partial S \subset S$ is the boundary of the simplex $S \subset P$. The homotopy $H$ maps into one of the open subsets of $B$ where $(E, p)$ is trivial up to simple maps, so we may replace $(E, p)$ by its restriction over this subset.

After this reduction, we may restrict to the absolute version of the homotopy lifting property again, because there is a homeomorphism of pairs

$$(S \times [a, b], S \times \{a\} \cup \partial S \times [a, b]) \cong (S \times [a, b], S \times \{a\}) .$$

Summarizing, we are only required to fill in the dashed arrow in (2.6.8) when $(E, p)$ is trivial up to a simple map.

Choose such a (proper, simple) trivialization $f : B \times F \to E$ over $B$, as in Definition 2.6.6. The simple map $f$ has the ALP by Proposition 2.6.5 ($E$ is metric). Therefore the map $g : P \to E$ has an approximate lifting $(\beta, \varphi) : P \times (0, 1] \to B \times F$ so that the union map $g \cup f(\beta, \varphi) : P \times [0, 1] \to E$ is continuous. Here we have written the approximate lifting as $(\beta, \varphi)$, in terms of its component maps $\beta : P \times (0, 1] \to B$ and $\varphi : P \times (0, 1] \to F$.

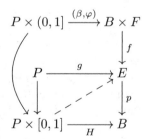

However, the map $g \cup f(\beta, \varphi)$ has no reason to be a lift over $p$ of the given homotopy $H$. Nonetheless this can be arranged, as follows.

Let $\eta : P \times (0, 1] \to B$ be the restriction of $H$ away from $P$, and replace $(\beta, \varphi)$ by the map $(\eta, \varphi) : P \times (0, 1] \to B \times F$. We claim that the union map

$g \cup f(\eta, \varphi) \colon P \times [0,1] \to E$ is continuous. It is then clear from its definition that it provides the dashed arrow in diagram (2.6.8), as required.

It suffices to prove continuity of $g \cup f(\eta, \varphi)$ at $(x, 0)$ in $P \times [0, 1]$ for each $x \in P$. Each neighborhood of the image point $g(x) \in E$ contains a compact neighborhood $N$ of $g(x)$, because $E$ is assumed to be locally compact. The preimage $f^{-1}(N)$ in $B \times F$ is compact, because $f$ is proper, so its projection $K = pr_2 f^{-1}(N) \subset F$ is also compact.

Assume for a contradiction that there is no neighborhood of $(x, 0) \in P \times [0, 1]$ that is mapped into $N$ by the union of $g$ and $f(\eta, \varphi)$. Then there is a sequence of points $(x_n, t_n)$ in $P \times [0, 1]$ converging to $(x, 0)$, whose images in $E$ are never in $N$. By the continuity of $g$ we may assume that none of the $t_n = 0$, so $(x_n, t_n)$ is a sequence of points in $P \times (0, 1]$, with $f(\eta, \varphi)(x_n, t_n)$ not in $N$.

Momentarily switching back from $(\eta, \varphi)$ to $(\beta, \varphi)$, by continuity of the map $P \times [0, 1] \to E$ in the ALP the sequence $f(\beta, \varphi)(x_n, t_n)$ converges to $g(x)$, hence eventually lies within $N$. Passing to a subsequence we may assume that all $(\beta, \varphi)(x_n, t_n)$ lie within $f^{-1}(N)$, so all $\varphi(x_n, t_n)$ lie within $K$. The latter space is compact, so by passing to a subsequence once more we may assume that $\varphi(x_n, t_n)$ converges to a point $y \in K$.

We also know that $\beta(x_n, t_n)$ and $\eta(x_n, t_n)$ converge to the same point $b = p(g(x))$ in $B$, by the continuity of the map in the ALP and the homotopy $H$, respectively. Thus $(\beta, \varphi)(x_n, t_n)$ and $(\eta, \varphi)(x_n, t_n)$ both converge to $(b, y)$ in $B \times F$, and so their image sequences under $f$ have the same limit in $E$. (These limits are unique, because $E$ is metric.) In the first case this limit is $g(x)$, by the continuity of the map in the ALP. In the second case $f(\eta, \varphi)(x_n, t_n)$ never enters the neighborhood $N$. This contradiction proves the claim of continuity for the union map $g \cup f(\eta, \varphi)$. $\square$

For a map of simplicial sets $\pi \colon Z \to \Delta^q$ we shall obtain a converse to Proposition 2.6.7 in Proposition 2.7.6. For a PL map $p \colon E \to |\Delta^q|$, we obtain the converse in Proposition 2.7.7.

**Definition 2.6.9.** We say that a map of simplicial sets is a **Serre fibration** if its geometric realization is a Serre fibration in the topological category, i.e., if it has the homotopy lifting property for continuous maps from compact polyhedra.

The following result applies to the cylinder reduction map $T(f) \to M(f)$, from the ordinary mapping cylinder to the reduced mapping cylinder, in the case of simple cylinder reduction. Here $T(f)$ and $M(f)$ are viewed as simplicial sets over $\Delta^1$ by means of the cylinder coordinate projections $\pi$, and as simplicial sets over $Y$ by means of the cylinder projections $pr$, see Definition 2.4.5. That special case will suffice for the proof of Proposition 2.7.6 for non-singular $Z$, while for singular $Z$ the following more general statement will be needed.

**Proposition 2.6.10.** *Let $f \colon X \to Y$ be a map of finite simplicial sets, and*

*suppose given a commutative diagram*

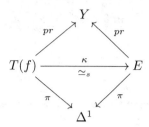

*of finite simplicial sets, where* $\kappa$ *is a simple map. Then* $f$ *is a simple map if and only if* $\pi\colon E \to \Delta^1$ *is a Serre fibration.*

*Proof.* Suppose first that $f$ is simple. Then the canonical map $|X| \times |\Delta^1| \to |T(f)|$ is simple by the gluing lemma. Hence its composite with the simple map $|\kappa|\colon |T(f)| \to |E|$ provides a trivialization up to a simple map of $|\pi|\colon |E| \to |\Delta^1|$, so $\pi\colon E \to \Delta^1$ is a Serre fibration by Proposition 2.6.7.

Conversely, suppose that $|\pi|\colon |E| \to |\Delta^1| \cong [0,1]$ is a Serre fibration. We shall show that $|f|\colon |X| \to |Y|$ has the ALP, so that $f$ is a simple map by Lemma 2.6.4.

Let $(P, Q)$ be a pair of compact polyhedra, and consider any maps $h\colon Q \to |X|$ and $g\colon P \to |Y|$ with $|f|h = g|Q$, as in Definition 2.6.1. These induce a map

$$h' = h \times id \cup g\colon Q \times [0,1] \cup_Q P \to |X| \times [0,1] \cup_{|X|} |Y| \cong |T(f)|$$

of topological mapping cylinders. Let $Q' = Q \times [0,1] \cup_Q P$ and $P' = P \times [0,1]$. Then $(P', Q')$ is another pair of compact polyhedra, and we obtain a commutative diagram

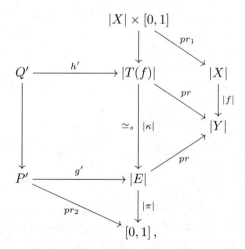

where $g'$ is a lift of $pr_2$ that extends $|\kappa|h'$, which exists because $|\pi|$ is a Serre fibration.

For the sake of motivation, we first consider the special case when $T(f) = E$ and $\kappa$ is the identity map. We can restrict $g' \colon P' = P \times [0,1] \to |E| = |T(f)|$ to the part over $(0,1] \subset [0,1]$, and form the composite map

$$P \times (0,1] \xrightarrow{g'|} |X| \times (0,1] \xrightarrow{pr_1} |X|,$$

where $|X| \times (0,1]$ is the part of $|T(f)|$ over $(0,1]$. On the subspace $Q \times (0,1] \subset Q'$, this map $pr_1 \circ g'|$ agrees with the composite map

$$Q \times (0,1] \xrightarrow{h \times id} |X| \times (0,1] \xrightarrow{pr_1} |X|,$$

which equals the restriction of $h \circ pr_1 \colon Q \times [0,1] \to |X|$ to the same subspace. Hence the union of $h \circ pr_1$ and $pr_1 \circ g'|$ define a map

$$Q \times \{0\} \cup P \times (0,1] \to |X|,$$

which, together with the composite map $pr \circ g' \colon P \times [0,1] \to |Y|$, verifies the approximate lifting property for $|f|$. The case where $\kappa$ is an isomorphism is similar.

Returning to the general case, we know that the simple map $|\kappa| \colon |T(f)| \to |E|$ has the relative ALP by Proposition 2.6.5. We apply this for the polyhedral pair $(P', Q')$ and the maps $h' \colon Q' \to |T(f)|$ and $g' \colon P' \to |E|$, to obtain maps $H$ and $G$ making the triangular prism in the following diagram commute.

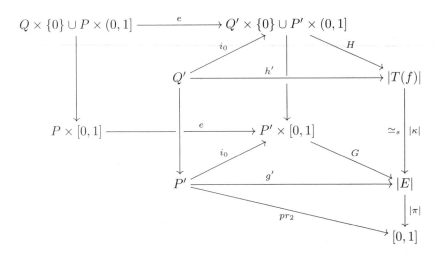

Here $i_0 \colon (p,t) \mapsto (p,t,0)$, for $(p,t) \in P'$, and similarly for $(p,t) \in Q'$. We may furthermore assume that the lifting $H$ is constant in the interval coordinate on the subpolyhedron $Q' \times [0,1]$, meaning that $H|Q' \times [0,1]$ equals the composite map

$$h' \circ pr_1 \colon Q' \times [0,1] \xrightarrow{pr_1} Q' \xrightarrow{h'} |T(f)|.$$

The part of $P' \times [0,1]$ over $(0,1]$ is an open subset

$$V = (|\pi|G)^{-1}((0,1]) \subset P \times [0,1] \times [0,1],$$

which contains $P \times (0,1] \times \{0\}$. Using the compactness of $P$, we can find a continuous function $\epsilon \colon [0,1] \to [0,1]$ with $\epsilon(0) = 0$ and $\epsilon(t) > 0$ for $t > 0$, such that $(p,t,\epsilon(t)) \in V$ for all $p \in P$ and $t > 0$.

Let $e \colon P \times [0,1] \to P' \times [0,1]$ be the embedding given by $e(p,t) = (p,t,\epsilon(t))$. The same formula defines a compatible embedding $e \colon Q \times \{0\} \cup P \times (0,1] \to Q' \times \{0\} \cup P' \times (0,1]$, making the left hand rectangle commute. Then the composite $He$ maps $P \times (0,1]$ to the part $|X| \times (0,1]$ of $|T(f)|$ over $(0,1]$, because $e(p,t) \in V$ for $t > 0$. Let $H' = pr_1 \circ He|$ be the composite map

$$H' \colon P \times (0,1] \xrightarrow{He|} |X| \times (0,1] \xrightarrow{pr_1} |X|.$$

The restriction $H'|Q \times (0,1]$ equals the composite map

$$Q \times (0,1] \xrightarrow{h'|} |X| \times (0,1] \xrightarrow{pr_1} |X|,$$

because $e(Q \times (0,1])$ lies in $Q' \times (0,1]$, where $H$ factors as $h' \circ pr_1$, and $pr_1 \circ e \colon Q \times (0,1] \to Q'$ is the inclusion. It is therefore also equal to the composite map

$$Q \times (0,1] \xrightarrow{pr_1} Q \xrightarrow{h} |X|,$$

in view of the definition of $h'$. Hence $H'$ and $h \circ pr_1 \colon Q \times [0,1] \to |X|$ agree on the subspace $Q \times (0,1]$, and their union defines a map

$$Q \times \{0\} \cup P \times (0,1] \to |X|,$$

which, together with the composite map

$$P \times [0,1] \xrightarrow{e} P' \times [0,1] \xrightarrow{G} |E| \xrightarrow{pr} |Y|,$$

verifies the ALP for $|f| \colon |X| \to |Y|$.  $\square$

## 2.7. Subdivision of simplicial sets over $\Delta^q$

In Section 3.1 we shall study simplicial categories whose objects in simplicial degree $q$ can be thought of as suitably continuous $q$-parameter families of simplicial sets. More precisely, the objects will be Serre fibrations $\pi \colon Z \to \Delta^q$ of simplicial sets. Recall that we say that a map of simplicial sets is a Serre fibration if its geometric realization is a Serre fibration (of topological spaces).

The purpose of the present section is to use the iterated reduced mapping cylinder of Section 2.4 to obtain combinatorial control of the property that $\pi \colon Z \to \Delta^q$ is a Serre fibration, in terms of simple maps. In Proposition 2.7.2 we recognize the part of $Sd(Z)$ sitting over a given simplex in the target of the subdivided map $Sd(\pi): Sd(Z) \to Sd(\Delta^q)$ as a naturally occurring iterated reduced mapping cylinder, at least when $Z$ is non-singular. This leads to the construction in Proposition 2.7.5 of a natural map $t$ to $Z$ from a product bundle over $\Delta^q$, and the study of the approximate lifting property from Section 2.6 lets us conclude in Proposition 2.7.6 that $\pi$ is a Serre fibration if and only if $t$ is a simple map. For use in Section 3.3, we prove a similar statement for PL Serre fibrations (of compact polyhedra) in Proposition 2.7.7.

We prove "fiber" and "base" gluing lemmas for Serre fibrations of simplicial sets in Lemmas 2.7.10 and 2.7.12, respectively.

**Definition 2.7.1.** Let $\pi: Z \to \Delta^q$ be any map of simplicial sets. Each face $\mu$ of $\Delta^q$ corresponds to a vertex $(\mu)$ of $Sd(\Delta^q)$. The preimage $Sd(\pi)^{-1}(\mu)$ of this vertex under the subdivided map $Sd(\pi): Sd(Z) \to Sd(\Delta^q)$ is a simplicial subset of $Sd(Z)$, which we call the **preimage of the barycenter of the face** $\mu$. The name may be justified, to some extent, by passage to geometric realization.

In the special case when $\mu = \iota_q = \beta$ (for barycenter) is the maximal face of $\Delta^q$, i.e., the one of dimension $q$, we simply call the simplicial set $Sd(\pi)^{-1}(\beta)$ the **preimage of the barycenter** associated to $\pi: Z \to \Delta^q$. When $q = 0$, $Sd(\pi)^{-1}(\beta) = Sd(Z)$.

**Proposition 2.7.2.** *Let $\pi: Z \to \Delta^q$ be any map of simplicial sets.*

*(a) Let $\mu \leq \nu$ be two comparable faces of $\Delta^q$, or what is the same, two vertices of $Sd(\Delta^q)$ that are connected by an edge. Then there is a natural map*

$$g = g_{\mu\nu}: Sd(\pi)^{-1}(\nu) \to Sd(\pi)^{-1}(\mu)$$

*(which we call the **pullback map** along that edge) between the preimages of the barycenters of these faces.*

*(b) Let $\lambda \leq \mu \leq \nu$ be comparable faces of $\Delta^q$. Then the identity*

$$g_{\lambda\nu} = g_{\lambda\mu} \circ g_{\mu\nu}: Sd(\pi)^{-1}(\nu) \to Sd(\pi)^{-1}(\lambda)$$

*holds.*

*(c) Suppose, furthermore, that $Z$ is non-singular, and let $s = (\mu_0 \leq \cdots \leq \mu_r)$ be an $r$-simplex in $Sd(\Delta^q)$. Then the part of $Sd(Z)$ over $s$, i.e., the fiber product of $Sd(\pi): Sd(Z) \to Sd(\Delta^q)$ and $\bar{s}: \Delta^r \to Sd(\Delta^q)$, is naturally isomorphic to the $r$-fold iterated backward reduced mapping cylinder of the resulting chain of pullback maps*

$$Sd(\pi)^{-1}(\mu_r) \to \cdots \to Sd(\pi)^{-1}(\mu_0).$$

*Proof.* (a) It suffices to construct the natural pullback map in the local case, when $Z = \Delta^n$ is an $n$-simplex for some $n \geq 0$, because the preimage of the barycenter functors commute with colimits. However, the same construction works for all non-singular $Z$, and this generality will be convenient in the proof of Lemma 2.7.4. We therefore only assume that $Z$ is non-singular, in which case $Sd(Z) = B(Z) = N(Z^\#)$.

By hypothesis, $\mu\colon [k] \to [q]$ is a face of $\nu\colon [\ell] \to [q]$, so there is a unique face operator $\lambda\colon [k] \to [\ell]$ such that $\mu = \nu\lambda$. The preimage $Sd(\pi)^{-1}(\nu)$ equals the nerve $NC$ of the partially ordered subset $C \subset Z^{\#}$ of non-degenerate simplices $v \in Z_n$ for $n \geq 0$ such that $\pi(v)^{\#} = \nu$, i.e., such that $\pi \circ \bar{v}$ factors as $\nu \circ \rho$ for some degeneracy operator $\rho$. Similarly, $Sd(\pi)^{-1}(\mu)$ equals $ND$ where $D \subset Z^{\#}$ consists of the non-degenerate simplices $w \in Z_m$ for $m \geq 0$ such that $\pi(w)^{\#} = \mu$.

Given $v \in C$, let the face operator $\lambda_1\colon [m] \to [n]$ and the degeneracy operator $\rho_1\colon [m] \to [k]$ be the pullbacks of $\lambda$ and $\rho$, as in the following diagram:

(2.7.3)

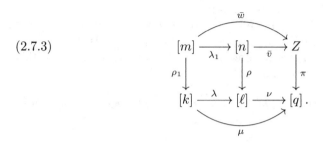

There is then an order-preserving function

$$\psi\colon C \to D$$
$$v \mapsto w = \lambda_1^*(v)$$

given by pullback along $\lambda$. Note that $w = \lambda_1^*(v)$ is already non-degenerate, because $Z$ is non-singular. The natural pullback map in question is defined as the nerve map $g = N\psi\colon NC \to ND$.

In other words, when $Z$ is non-singular, $g$ is a map of nerves and is determined by its effect on vertices. A vertex of $Sd(\pi)^{-1}(\nu)$ is a non-degenerate simplex $v$ of $Z$ that maps onto $\nu$ in $\Delta^q$, meaning that $\nu$ is the non-degenerate part of $\pi(v)$. The maximal face $w$ of $v$ that lies over the face $\mu$ of $\nu$ is then a non-degenerate simplex of $Z$ and represents a vertex of $Sd(\pi)^{-1}(\mu)$. The pullback map $g$ takes $v$ to this vertex $w$.

(b) This is clear from the construction in (a).

(c) We now assume that $Z$ is non-singular. (The claim does not in general hold for singular $Z$.) At first, we shall also assume that the simplex $s = (\mu_0 < \cdots < \mu_r)$ is non-degenerate, because this makes the following argument a little clearer.

The part of $Sd(Z) = N(Z^{\#})$ over $s$ is the nerve of the partially ordered subset $P$ of $Z^{\#}$ that consists of the non-degenerate simplices $v \in Z_n$, $n \geq 0$, such that the image of $\pi \circ \bar{v}$ equals the image of one of the $\mu_0, \ldots, \mu_r$. The $\mu_j$ are all different, because $s$ is non-degenerate, so this partially ordered subset is a disjoint union $P = D_r \sqcup \cdots \sqcup D_0$, where $D_j$ is the set of $v \in Z^{\#}$ with $\pi(v)^{\#} = \mu_j$, for each $j \in [r]$. The subset partial ordering on $P$ is such that

$v > w$ for $v \in D_j$, $w \in D_{j-1}$ if and only if $w$ is a face of the image $\psi(v)$ of $v$ in $D_{j-1}$, due to the universal property of pullbacks. Thus

$$P \cong P(D_r \to \cdots \to D_0)$$

is an isomorphism of partially ordered sets, inducing the claimed isomorphism

$$NP \cong M(ND_r \to \cdots \to ND_0)$$

of simplicial sets, by Lemma 2.4.14.

Finally, any simplex of $Sd(\Delta^q)$ has the form $\alpha^*(s) = (\mu_{\alpha(0)} \leq \cdots \leq \mu_{\alpha(p)})$ for a degeneracy operator $\alpha \colon [p] \to [r]$ and a non-degenerate $r$-simplex $s = (\mu_0 < \cdots < \mu_r)$, as before. By definition, the part of $Sd(Z)$ over $\alpha^*(s)$ is the pullback along $\alpha$ of the part of $Sd(Z)$ over $s$. Since nerves commute with pullbacks, this is the same as the nerve of the pullback

$$
\begin{array}{ccc}
D_{\alpha(0)} \sqcup \cdots \sqcup D_{\alpha(p)} & \longrightarrow & D_0 \sqcup \cdots \sqcup D_r \\
\downarrow & & \downarrow \\
[p] & \xrightarrow{\;\;\;\alpha\;\;\;} & [r] \, ,
\end{array}
$$

i.e., the $p$-fold iterated backward reduced mapping cylinder of the chain of pullback maps

$$Sd(\pi)^{-1}(\mu_{\alpha(p)}) \to \cdots \to Sd(\pi)^{-1}(\mu_{\alpha(0)}) \, .$$

$\square$

**Lemma 2.7.4.** *Let $\pi \colon Z \to \Delta^q$ be a map of finite non-singular simplicial sets, and let $\mu \leq \nu$ be faces of $\Delta^q$. Then the natural pullback map*

$$g \colon Sd(\pi)^{-1}(\nu) \to Sd(\pi)^{-1}(\mu)$$

*has simple backward cylinder reduction $T(g) \to M(g)$.*

See Lemma 2.7.9 for a related result.

*Proof.* We keep the notations $C$, $D$, $\psi$ from the proof of Proposition 2.7.2(a), so that $Sd(\pi)^{-1}(\nu) = NC$, $Sd(\pi)^{-1}(\mu) = ND$ and $g = N\psi$. We shall verify that $N\psi \colon N(C/v) \to N(D/w)$ is a simple map, for each $v \in C$ and $w = \psi(v) \in D$. The conclusion of the lemma then follows by Proposition 2.4.16.

An element $v \in C$ is a non-degenerate simplex $v \in Z_n$ such that $\pi \circ \bar{v} = \nu \circ \rho$ for some degeneracy operator $\rho$. See (2.7.3). The left ideal $C/v$ consists of the faces $\nu_1^*(v)$ of $v$, for $\nu_1 \colon [n_1] \to [n]$ injective, such that $\rho\nu_1$ remains surjective. Its nerve

$$N(C/v) = \prod_{i \in [\ell]} N(\rho^{-1}(i))$$

is the product of the simplices spanned by the (non-empty) preimages of $\rho$.

The image $w = \psi(v) = \lambda_1^*(v) \in D$ is a non-degenerate simplex $w \in Z_m$ such that $\pi \circ \bar{w} = \mu \circ \rho_1$. The left ideal $D/w$ consists of the faces $\mu_1^*(w)$ of $w$, for $\mu_1 \colon [m_1] \to [m]$ injective, such that $\rho_1\mu_1$ remains surjective. Its nerve

$$N(D/w) = \prod_{j \in [k]} N(\rho_1^{-1}(j))$$

is the product of the simplices spanned by the (non-empty) preimages of $\rho_1$.

Furthermore, for each $j$ the map $\lambda_1$ identifies $\rho_1^{-1}(j) \subset [m]$ with $\rho^{-1}(\lambda(j)) \subset [n]$, so that the map of nerves $N\psi \colon N(C/v) \to N(D/w)$ is the projection from the product indexed over $i \in [\ell]$ onto the subset of factors indexed over $j \in [k]$. In other words, this is the projection away from the factors indexed by the $i$ not in the image of $\lambda$. Hence each point inverse after geometric realization is a product of simplices, and therefore contractible. $\square$

Recall from Definition 2.7.1 that $\beta = \iota_q$ denotes the maximal face of $\Delta^q$.

**Proposition 2.7.5.** *There is a natural transformation (of endofunctors of the category of simplicial sets over $\Delta^q$) which to an object $\pi \colon Z \to \Delta^q$ associates a map*

$$t \colon Sd(\pi)^{-1}(\beta) \times \Delta^q \to Z$$

*over $\Delta^q$ with the following two properties:*
*(a) The map $t$ is simple over the interior of $|\Delta^q|$;*
*(b) For each face $\mu \colon \Delta^p \to \Delta^q$, the diagram*

$$
\begin{array}{ccccc}
Sd(\pi)^{-1}(\beta) \times \Delta^p & \xrightarrow{id \times \mu} & Sd(\pi)^{-1}(\beta) \times \Delta^q & \xrightarrow{\;\;t\;\;} & Z \\
{\scriptstyle g \times id}\downarrow & & & & \uparrow \\
Sd(\pi)^{-1}(\mu) \times \Delta^p & & \xrightarrow{\hspace{4cm}t'\hspace{4cm}} & & Z \times_{\Delta^q} \Delta^p
\end{array}
$$

*commutes.*

In (a), "simple over the interior" is in the sense of Remark 2.1.9, i.e., the preimage $|t|^{-1}(z)$ is contractible for each point $z \in |Z|$ with $|\pi|(z)$ in the interior of $|\Delta^q|$. In (b), $t'$ is the natural map for the restricted map $\pi \colon Z \times_{\Delta^q} \Delta^p \to \Delta^p$, and $g \colon Sd(\pi)^{-1}(\beta) \to Sd(\pi)^{-1}(\mu)$ is the pullback map of Proposition 2.7.2(a). Property (a) describes $t$ over the interior of $|\Delta^q|$, while (b) describes it over the boundary. When $q = 0$, the map $t$ will be equal to the last vertex map $d_Z \colon Sd(Z) \to Z$.

*Proof.* As the preimage of the barycenter functor commutes with pushouts, the construction of the natural map $t$ only needs to be given in the local case where $Z = \Delta^n$ is a simplex. Properties (a) and (b) also only need to be verified in that case. This is clear for (b), and uses the gluing lemma for simple maps in its modified version (Remark 2.1.9) for property (a). So assume from now on that $\pi = N\varphi \colon \Delta^n \to \Delta^q$ is induced by an order-preserving function $\varphi \colon [n] \to [q]$.

If $\varphi$ is not surjective, then $Sd(\pi)^{-1}(\beta)$ is empty, so the asserted natural map $t: \emptyset \to \Delta^n$ exists for trivial reasons. The asserted simplicity over the interior of $|\Delta^q|$ is also a trivial fact, for $\pi$ maps $\Delta^n$ entirely into the boundary $\partial \Delta^q$.

We can therefore assume that $\varphi: [n] \to [q]$ is a surjective map. The transformation $t$ to be constructed will be a simple map in this case, not just a simple map over the interior of $|\Delta^q|$. For each $j = 0, \dots, q$ let $E_j = \varphi^{-1}(j) \subset [n]$. Then each $E_j$ is a non-empty totally ordered set, $NE_j$ is a simplex, and $[n] = E_0 \sqcup \cdots \sqcup E_q$ is a decomposition as a disjoint union.

There is a natural isomorphism

$$Sd(\pi)^{-1}(\beta) \cong Sd(NE_0) \times \cdots \times Sd(NE_q).$$

Here $Sd(\pi)^{-1}(\beta) = NC$ is the nerve of the partially ordered set $C$ of (non-empty) faces of $\Delta^n$ that map onto $\Delta^q$ by $\pi$, i.e., the subsets of $[n]$ that intersect each $E_j$ nontrivially. We naturally identify $C$ with the product over $j = 0, \dots, q$ of the partially ordered set of non-empty subsets of $E_j$, i.e., of (non-empty) faces of the simplex $NE_j$. The nerve of each factor is the Barratt nerve of $NE_j$, which is isomorphic to $Sd(NE_j)$. Thus $NC$ is isomorphic to the nerve of the product, which is the product just displayed.

For each $j$ there is a last vertex map $d = d_{NE_j}: Sd(NE_j) \to NE_j$, which is simple by Proposition 2.2.18. Hence their product

$$d^q: Sd(NE_0) \times \cdots \times Sd(NE_q) \to NE_0 \times \cdots \times NE_q$$

is also a simple, natural map. We shall in a moment construct a natural map

$$N\tau: NE_0 \times \cdots \times NE_q \times \Delta^q \to \Delta^n$$

over $\Delta^q$, which is a simplicial homotopy equivalence over the target (as in Definition 2.4.7), and therefore simple. The composite map

$$t: Sd(\pi)^{-1}(\beta) \times \Delta^q \cong Sd(NE_0) \times \cdots \times Sd(NE_q) \times \Delta^q$$
$$\xrightarrow{d^q \times id} NE_0 \times \cdots \times NE_q \times \Delta^q \xrightarrow{N\tau} \Delta^n$$

will then be the required simple map $t$ over $\Delta^q$.

We define $N\tau$ as the nerve of the order-preserving function

$$\tau: E_0 \times \cdots \times E_q \times [q] \to [n]$$

given by

$$\tau(i_0, \dots, i_q, j) = i_j.$$

Here each $i_j \in E_j \subset [n]$.

We define an order-preserving section $\sigma: [n] \to E_0 \times \cdots \times E_q \times [q]$ by

$$\sigma(i_j) = (\max E_0, \dots, \max E_{j-1}, i_j, \min E_{j+1}, \dots \min E_q, j).$$

(The minima and maxima exist because each $E_j$ is non-empty.) Note that $\tau\sigma = id_{[n]}$. There is a chain of two natural inequalities relating the other composite $\sigma\tau$ and the identity on $E_0 \times \cdots \times E_q \times [q]$, as follows:

$$(i_0, \ldots, i_q, j) \leq (\max E_0, \ldots, \max E_{j-1}, i_j, i_{j+1}, \ldots, i_q, j)$$
$$\geq (\max E_0, \ldots, \max E_{j-1}, i_j, \min E_{j+1}, \ldots, \min E_q, j) \,.$$

Passing to nerves, these define a chain of two simplicial homotopies over $\Delta^n$ from $id$ to $N\sigma \circ N\tau$. Thus $N\tau$ is a simplicial homotopy equivalence over the target, and in particular, a simple map.

From $\varphi(i_j) = j$ we see that $\tau$ is a function over $[q]$, so $N\tau$ and $t$ are maps over $\Delta^q$. A morphism from $\varphi \colon [n] \to [q]$ to $\varphi' \colon [n'] \to [q]$ defines functions $E_j \to E_j'$, and $\tau$ is natural with respect to these. For each face operator $\mu \colon [p] \to [q]$ the pullback map $g$ is compatible under the last vertex maps with the projection from $E_0 \times \cdots \times E_q$ to the factors corresponding to the image of $\mu$. A review of the definitions therefore shows that the map $t$ satisfies properties (a) and (b), and is natural in $\pi = N\varphi$. This completes the proof of the proposition. $\square$

**Proposition 2.7.6.** *Let $\pi \colon Z \to \Delta^q$ be a map of finite simplicial sets. Then the following conditions are equivalent:*
*(a) $\pi$ is a Serre fibration;*
*(b) The pullback map*

$$g \colon Sd(\pi)^{-1}(\beta) \to Sd(\pi)^{-1}(\mu)$$

*is simple, for each face $\mu$ of the maximal face $\beta$ of $\Delta^q$;*
*(c) The natural map*

$$t \colon Sd(\pi)^{-1}(\beta) \times \Delta^q \to Z$$

*over $\Delta^q$ is simple.*

*Proof.* Consider an edge $s = (\mu \leq \beta)$ in $Sd(\Delta^q)$ with representing map $\bar{s} \colon \Delta^1 \to Sd(\Delta^q)$, and associate to each $\pi \colon Z \to \Delta^q$ the part

$$E(Z) = \Delta^1 \times_{Sd(\Delta^q)} Sd(Z)$$

of $Sd(Z)$ over that edge. Let $g_Z \colon Sd(\pi)^{-1}(\beta) \to Sd(\pi)^{-1}(\mu)$ be the usual pullback map over $s$. For $\pi \colon D \to \Delta^q$ with $D$ non-singular, we have natural maps

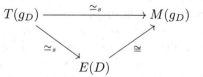

over $\Delta^1$ and $Sd(\pi)^{-1}(\mu)$, where the lower right hand map is an isomorphism by Proposition 2.7.2(c), and the upper map, the cylinder reduction map, is simple by Lemma 2.7.4. The lower left hand map is defined to make the diagram commute.

For each object $([n], z)$ of the simplex category of $Z$, the map $\pi\bar{z}\colon \Delta^n \to \Delta^q$ makes $D = \Delta^n$ a non-singular simplicial set over $\Delta^q$, so we can form the colimit over $\mathrm{simp}(Z)$ of the objects and maps in the previous triangular diagram. The constructions $Z \mapsto T(g_Z)$ and $Z \mapsto E(Z)$ commute with such colimits, and give us natural maps

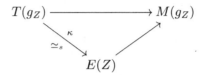

over $\Delta^1$ and $Sd(\pi)^{-1}(\mu)$. The construction $Z \mapsto M(g_Z)$ does not in general commute with colimits, so the lower right hand map will not in general be an isomorphism. The upper map is still the cylinder reduction map.

We claim that the lower left hand map, $\kappa\colon T(g_Z) \to E(Z)$, is simple. The constructions $Z \mapsto T(g_Z)$ and $Z \mapsto E(Z)$ both preserve pushout diagrams and cofibrations, so this follows from the non-degenerate case by inductively writing $Z = \Delta^n \cup_{\partial\Delta^n} Z'$, and an appeal to the gluing lemma for simple maps.

Applying Proposition 2.6.10, we have proved that $g = g_Z\colon Sd(\pi)^{-1}(\beta) \to Sd(\pi)^{-1}(\mu)$ is a simple map if and only if $\pi\colon E(Z) \to \Delta^1$ is a Serre fibration. We now use this to establish the equivalence of the three stated conditions.

Suppose first that $\pi\colon Z \to \Delta^q$ is a Serre fibration. In Theorem 2.3.2 we showed that the homeomorphism $h_Z\colon |Sd(Z)| \cong |Z|$ of Fritsch and Puppe [FP67] can be adjusted to a quasi-natural homeomorphism $h^\pi$, so as to make the square

$$
\begin{array}{ccc}
|Sd(Z)| & \xrightarrow[\cong]{h^\pi} & |Z| \\
{\scriptstyle |Sd(\pi)|}\Big\downarrow & & \Big\downarrow{\scriptstyle |\pi|} \\
|Sd(\Delta^q)| & \xrightarrow[\cong]{h_q} & |\Delta^q|
\end{array}
$$

commute. It results that the map $Sd(\pi)\colon Sd(Z) \to Sd(\Delta^q)$ is a Serre fibration, because this is a topological property of its geometric realization. Consequently, for each face $\mu$ of $\Delta^q$ the restriction $\pi\colon E(Z) \to \Delta^1$ of $Sd(\pi)$ to the part over $(\mu \le \beta)$ in $Sd(\Delta^q)$ is also a Serre fibration. As we have just summarized, this implies that the pullback map $g$ is simple.

Next, suppose that $g$ is simple for each face $\mu$ of $\Delta^q$. Then the left hand vertical map $g \times id$ in the diagram in property (b) of Proposition 2.7.5 is simple. The bottom map $t'$ is simple over the interior of $\Delta^p$, in view of property (a) of that proposition. Using the commutativity of the diagram, we see therefore that the map $t$ is simple over the interior of the face $\mu(\Delta^p)$ of $\Delta^q$. But that face was arbitrary. Hence $t$ is simple.

Finally, if $t$ is simple, then its geometric realization provides a trivialization of $|\pi| \colon |Z| \to |\Delta^q|$ up to a simple map, so $\pi \colon Z \to \Delta^q$ is a Serre fibration by Proposition 2.6.7. $\square$

We shall also need a PL version of this simplicial result. We say that a finite non-singular simplicial set $X$ is **simplicially collapsible** if for some vertex $v$ of $X$ there is a simplicial expansion $\{v\} \subset X$, cf. Definition 3.2.1(d). By [Wh39, Thms. 6 and 7], any triangulation of $|\Delta^q|$ admits a subdivision that is simplicially collapsible. Hence any PL map $E \to |\Delta^q|$ admits a triangulation $Y \to X$ with $X$ simplicially collapsible.

**Proposition 2.7.7.** *Each PL Serre fibration $p \colon E \to |\Delta^q|$ can be trivialized up to a simple map.*

*More precisely, given a triangulation $\pi \colon Y \to X$ of $p$ by finite non-singular simplicial sets, with $X$ simplicially collapsible, there exists a functorially defined finite non-singular simplicial set $L_X$ and a natural simple map*

$$t_X \colon L_X \times X \to Y$$

*over $X$. Its polyhedral realization $|t_X| \colon |L_X| \times |\Delta^q| \to E$ over $|\Delta^q|$ then trivializes $p$ up to a simple map.*

*Proof.* For each non-degenerate simplex $x$ in $X$, let $\Phi(x) = Sd(\pi)^{-1}(\beta_x)$ be the preimage of the barycenter $\beta_x$ of that simplex. If $x$ is a face of another non-degenerate simplex $y$ in $X$, so $x \leq y$ in $X^\#$, then there is a natural pullback map $g \colon \Phi(y) \to \Phi(x)$, as in Proposition 2.7.2(a). By Proposition 2.7.2(b) it makes sense to form the limit

$$L_X = \lim_{x \in (X^\#)^{op}} \Phi(x).$$

It is a finite non-singular simplicial set, because it is contained in the corresponding finite product of finite non-singular simplicial sets $\Phi(x) \subset Sd(Y)$. This construction is clearly functorial in (finite, non-singular) simplicial sets over $X$.

Let $pr_x \colon L_X \to \Phi(x)$ be the canonical projection from the limit. We shall prove in Lemma 2.7.8 that each $pr_x$ is a simple map. Assuming this, we can construct the desired simple map $t_X \colon L_X \times X \to Y$ over $X$ by gluing together the composite maps

$$L_X \times \Delta^m \xrightarrow{pr_x \times id} \Phi(x) \times \Delta^m \xrightarrow{t_x} Y \times_X \Delta^m \subset Y,$$

for all non-degenerate $x$ in $X$. Here $Z = Y \times_X \Delta^m \to \Delta^m$ is the pullback of $\pi \colon Y \to X$ along the representing map $\bar{x} \colon \Delta^m \to X$, and $t = t_x \colon \Phi(x) \times \Delta^m \to Z$ is the natural map of Proposition 2.7.5. By hypothesis $Y \to X$ is a Serre fibration, so $Z \to \Delta^m$ is also a Serre fibration, and $t_x$ is a simple map by Proposition 2.7.6. The naturality of $t_X$ follows from the naturality of the maps $t_x$.

To see that these maps glue together, we need to know that the outer rectangle of the following diagram commutes whenever $x \leq y$ in $X^{\#}$, so that $x = \mu^*(y)$ for some face operator $\mu\colon [m] \to [n]$.

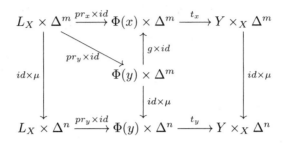

The triangle commutes by the definition of $L_X$ as a limit over the pullback maps $g$, the left hand quadrilateral obviously commutes, and the right hand part of the diagram commutes by Proposition 2.7.5(b). Hence the glued map $L_X \times X \to Y$ over $X$ is well-defined, and is simple by the gluing lemma. $\square$

**Lemma 2.7.8.** *Let $\pi\colon Y \to X$ be a Serre fibration of finite non-singular simplicial sets, with $X$ simplicially collapsible. Then each projection $pr_x\colon L_X \to \Phi(x) = Sd(\pi)^{-1}(\beta_x)$ is a simple map.*

*Proof.* When $X$ is a point, there is nothing to prove. Otherwise, we may write $X = \Delta^n \cup_{\Lambda_i^n} X'$ where $\Lambda_i^n \subset \Delta^n$ is the $i$-th horn. We define $L_{\Delta^n}$, $L_{\Lambda_i^n}$ and $L_{X'}$ by analogy with $L_X$, for the restriction of $\pi\colon Y \to X$ to $\Delta^n$, $\Lambda_i^n$ and $X' \subset X$, respectively. By induction we may assume that $pr_x\colon L_{X'} \to \Phi(x)$ is a simple map for each $x \in (X')^{\#}$. Note also that if $x \leq y$ are the two non-degenerate simplices in $X$ that are not contained in $X'$, with $\bar{y}\colon \Delta^n \to X$ and $x$ the $i$-th face of $y$, then $pr_y\colon L_{\Delta^n} \to \Phi(y)$ is an isomorphism and $pr_x\colon L_{\Delta^n} \to \Phi(x)$ can be identified with the map $g\colon \Phi(y) \to \Phi(x)$. The isomorphism is obviously a simple map, and $g$ is a simple map by Proposition 2.7.6, because $\pi$ restricted to $\Delta^n$ is a Serre fibration. There is a pullback diagram of limits

$$
\begin{array}{ccc}
L_X & \longrightarrow & L_{X'} \\
\downarrow & & \downarrow \\
L_{\Delta^n} & \xrightarrow{\ f\ } & L_{\Lambda_i^n} \ .
\end{array}
$$

Given the observations above, it will suffice to show that $L_X \to L_{X'}$ and $L_X \to L_{\Delta^n}$ are simple maps. By the pullback property of simple maps (Proposition 2.1.3(c)) and a second induction, this will follow once we show that the lower horizontal map $f\colon L_{\Delta^n} \to L_{\Lambda_i^n}$ is simple. For $X$ can be obtained from $\Delta^n$ by filling finitely many embedded horns.

We need to introduce some notation. Let $\pi\colon Z = Y \times_X \Delta^n \to \Delta^n$ be the restriction of $\pi\colon Y \to X$ to $\Delta^n$. For each face $\mu$ of $\Delta^n$, let $Z_{\mu}^{\#} \subset Z^{\#}$ be the partially ordered set of non-degenerate simplices $z$ of $Z$ such that $\pi(z)^{\#} = \mu$, so $Sd(\pi)^{-1}(\mu) = NZ_\mu^\#$. Let $C = Z_\beta^\#$, where $\beta$ is the maximal non-degenerate face of $\Delta^n$, and let $D = \lim_{i\in\mu\neq\beta} Z_\mu^\#$, where $\mu$ ranges over the non-degenerate simplices of $\Lambda_i^n$, so $L_{\Delta^n} = NC$ and $L_{\Lambda_i^n} = ND$. The limit is formed over the order-preserving functions $\psi \colon Z_\nu^\# \to Z_\mu^\#$ with $g = N\psi$, for $\mu \leq \nu$. The canonical order-preserving function $\varphi \colon C \to D$ induces $f = N\varphi$ at the level of nerves.

We can then identify

$$Z_\beta^\# \sqcup_\varphi (\lim_{i\in\mu\neq\beta} Z_\mu^\#) \cong \lim_{i\in\mu\neq\beta} (Z_\beta^\# \sqcup_\psi Z_\mu^\#)$$

as partially ordered sets, where the limit on the right is formed over the order-preserving functions $id \sqcup \psi$. Passing to nerves, we obtain an isomorphism

$$M(f) = M(\Phi(\beta) \to \lim_{i\in\mu\neq\beta} \Phi(\mu)) \cong \lim_{i\in\mu\neq\beta} M(\Phi(\beta) \to \Phi(\mu))$$

of simplicial sets over $\Delta^1$. Hence we can identify the cylinder coordinate projection $M(f) \to \Delta^1$ with the limit of the cylinder coordinate projections $M(g) = M(\Phi(\beta) \to \Phi(\mu)) \to \Delta^1$. Each of the latter projections is a Serre fibration, by the hypothesis, so also their limit is a Serre fibration (this is so by the universal property defining limits because, in the case at hand, geometric realization preserves limits). Hence $f$ is a simple map, by the case $E = M(f)$ of Proposition 2.6.10, once we know that the cylinder reduction map $T(f) \to M(f)$ is simple. But that is the content of the following lemma, in view of Proposition 2.4.16. $\square$

**Lemma 2.7.9.** *For each $v \in C = Z_\beta^\#$, and $w = \varphi(v) \in D = \lim_{i\in\mu\neq\beta} Z_\mu^\#$, the limit map*

$$N\varphi \colon N(C/v) \to N(D/w)$$

*is simple.*

*Proof.* The composite $\rho = \pi \circ \bar{v}$ is a degeneracy operator with target $[n]$. Thus

$$N(C/v) = \prod_{j\in[n]} N(\rho^{-1}(j))$$

is a product of (non-empty) simplices. The image $\varphi(v) = w = (w_\mu)_{i\in\mu\neq\beta}$ is a compatible family of elements $w_\mu \in Z_\mu^\#$. Here $w_\mu$ is the face of $v$ spanned by the vertices that lie over $\mu$. So

$$N(Z_\mu^\#/w_\mu) \cong \prod_{j\in\mu} N(\rho^{-1}(j))$$

and $N(C/v) \to N(Z_\mu^\#/w_\mu)$ is the projection onto the factors indexed by the vertices of $\mu$. It follows that

$$N(D/w) = \lim_{i\in\mu\neq\beta} \prod_{j\in\mu} N(\rho^{-1}(j)) \cong \prod_j N(\rho^{-1}(j))$$

where now $j$ ranges over the vertices of $\Lambda_i^n$. When $n \geq 2$, the horn $\Lambda_i^n$ contains all the vertices of $\Delta^n$, so

$$N\varphi \colon N(C/v) \to N(D/w)$$

is an isomorphism, and thus simple. When $n = 1$, the horn is a single vertex, and $N\varphi$ is the projection from a product of two simplices to one of these simplices, which is also simple. $\square$

**Lemma 2.7.10 (Fiber gluing lemma for Serre fibrations).** *Let*

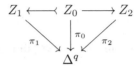

*be a commutative diagram of finite simplicial sets, where $\pi_i \colon Z_i \to \Delta^q$ is a Serre fibration for each $i = 0, 1, 2$, and $Z_0 \to Z_1$ is a cofibration. Then the induced map*

$$\pi \colon Z_1 \cup_{Z_0} Z_2 \to \Delta^q$$

*is a Serre fibration.*

*Proof.* By Proposition 2.7.6 we have simple maps $t_i \colon Sd(\pi_i)^{-1}(\beta) \times \Delta^q \to Z_i$, for $i = 0, 1, 2$. These combine, by naturality, to a map

$$t_1 \cup_{t_0} t_2 \colon \left( Sd(\pi_1)^{-1}(\beta) \cup_{Sd(\pi_0)^{-1}(\beta)} Sd(\pi_2)^{-1}(\beta) \right) \times \Delta^q \to Z_1 \cup_{Z_0} Z_2$$

which, by the gluing lemma, is also simple. We have thus obtained a trivialization of $\pi$ up to a simple map. Hence (Proposition 2.6.7) the map $\pi \colon Z_1 \cup_{Z_0} Z_2 \to \Delta^q$ is a Serre fibration.

For an alternative proof, we may appeal to results of M. Steinberger, J. West and M. Clapp. According to [SW84, Thm. 1], each Serre fibration $|\pi| \colon |Z_i| \to |\Delta^q|$ of CW complexes is a Hurewicz fibration in the convenient category of weak Hausdorff compactly generated spaces, for $i = 0, 1, 2$. Since $|Z_0| \to |Z_1|$ is a closed cofibration, it follows from [Cl81, Prop. 1.3] that the pushout $|Z_1 \cup_{Z_0} Z_2| \cong |Z_1| \cup_{|Z_0|} |Z_2| \to |\Delta^q|$ is a Hurewicz fibration, in the same category. Hence it is also a Serre fibration. $\square$

**Definition 2.7.11.** We say that a map $\pi \colon E \to B$ of simplicial sets is a **simplex-wise Serre fibration** if for each simplex $y$ of $B$, with representing map $\bar{y} \colon \Delta^q \to B$, the pulled-back map $\bar{y}^*\pi \colon \Delta^q \times_B E \to \Delta^q$ is a Serre fibration.

Pullbacks preserve Serre fibrations, so any Serre fibration is a simplex-wise Serre fibration, and if $\pi \colon E \to B$ pulls back to a Serre fibration over each non-degenerate simplex of $B$, then it is a simplex-wise Serre fibration. For non-singular $B$, this amounts to asking that $\pi$ restricts to a Serre fibration over each cofibration $\Delta^q \to B$, for $q \geq 0$.

**Lemma 2.7.12 (Base gluing lemma for Serre fibrations).** *Let $\pi\colon E \to B$ be a map of simplicial sets, with $B$ finite and non-singular. If $\pi$ is a simplex-wise Serre fibration then $\pi$ is a Serre fibration.*

*Proof.* It follows from Theorem 2.3.2 that if $\pi$ is a simplex-wise Serre fibration then $Sd(\pi)\colon Sd(E) \to Sd(B)$ is a simplex-wise Serre fibration, and that if $Sd(\pi)$ is a Serre fibration then $\pi$ is a Serre fibration. Hence it suffices to prove the lemma for $Sd(\pi)$. We may therefore assume, from the outset, that $\pi\colon E \to B$ is a cellular map of CW complexes, that $B$ is more specifically a finite simplicial complex, and that $\pi$ restricts to a Serre fibration $E_\Delta = \pi^{-1}(\Delta) \to \Delta$ over each simplex $\Delta \subset B$.

In order to prove that $\pi$ is a Serre fibration, it suffices by the argument after diagram (2.6.8) to find an open cover of $B$ such that $\pi$ restricts to a Serre fibration over each subset in that cover. We shall show that $\pi$ restricts to a Serre fibration over the star neighborhood $St(v, B)$ of each vertex in $B$ (see Definition 3.2.11). The required open cover is then given by the interiors of these stars, i.e., by the open star neighborhoods of the vertices in $B$.

To simplify the notation, we suppose that $B = St(v, B)$ is itself the star of a fixed vertex $v$. Then $B \cong \mathrm{cone}(L)$ is a cone with vertex $v$ and base the link $L = Lk(v, B)$. We can build $L$ by starting with the empty space and inductively attaching simplices along their boundaries. In the same way we can build $B$ by starting with $v$ and inductively attaching cones on simplices, which are simplices $\Delta$ containing $v$ as a vertex, along cones on their boundaries, which are horns $\Lambda$ consisting of the proper faces of $\Delta$ that contain $v$.

The strategy for the remainder of the proof will be to show that $\pi$ is a retract of the product fibration $pr\colon E \times B \to B$, from which it formally follows that $\pi$ is a Serre fibration. We start with the embedding

$$i = (id, \pi)\colon E \to E \times B$$

of $E$ as the graph of $\pi$. This is a cellular map of CW complexes over $B$. We will construct a left inverse to $i$ over $B$, i.e., a continuous map $r\colon E \times B \to E$ with $r \circ i = id$ and $\pi \circ r = pr$.

The construction of $r$ proceeds in two steps.

First, let $c\colon B \times I \to B$ be the linear contraction of $B$ to the cone point $v$, with $c(b, 1) = b$ and $c(b, 0) = v$ for $b \in B$, and let $f\colon E \times I \to B$ be $\pi \times id\colon E \times I \to B \times I$ composed with $c$, so that $f(e, t) = c(\pi(e), t)$. We claim that there is a lift $F\colon E \times I \to E$ of $f$ such that $F(e, 1) = e$ for $e \in E$.

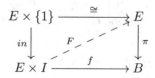

We shall prove this by an induction over the (positive-dimensional) simplices $\Delta$ of $B$ that contain $v$ as a vertex, in order of increasing dimension. To start the induction, let $E_v = \pi^{-1}(v)$ and define $F$ on $E_v \times I$ by $F(e, t) = e$. For the inductive step, consider such a simplex $\Delta$, and let $\Lambda$ be the union of the proper faces of $\Delta$ that contain $v$. Let $\pi_\Delta \colon E_\Delta \to \Delta$ and $E_\Lambda = \pi^{-1}(\Lambda) \to \Lambda$ be the restrictions of $\pi$ to these subcomplexes. The maps $c$ and $f$ are compatible with these restrictions. If $F$ has already been defined on $E_\Lambda \times I$, then the desired extension to $E_\Delta \times I$ is a lift in the following diagram.

$$
\begin{array}{ccc}
E_\Delta \times \{1\} \cup E_\Lambda \times I & \longrightarrow & E_\Delta \\
{\scriptstyle in} \downarrow & & \downarrow {\scriptstyle \pi_\Delta} \\
E_\Delta \times I & \longrightarrow & \Delta
\end{array}
$$

Here the upper horizontal map takes $(e, 1)$ to $e$, and is given by the previously defined $F$ on $E_\Lambda \times I$. The lower horizontal map is the restriction of $f$. Such a lift exists because $E_\Lambda$ is a CW subcomplex of $E_\Delta$ and $\pi_\Delta$ is a Serre fibration. By induction, it follows that $F$ can be extended to all of $E \times I$.

Second, let $g \colon E \times B \times I \to B$ be the projection to $B \times I$ composed with $c$, so that $g(e, b, t) = c(b, t)$. We claim that there is a lift $G \colon E \times B \times I \to E$ of $g$ such that $G(e, b, 0) = F(e, 0)$ for $e \in E$ and $b \in B$, and $G(i(e), t) = F(e, t)$ for $e \in E$ and $t \in I$. In other words, $G$ is a lift in the following diagram.

$$
\begin{array}{ccc}
E \times B \times \{0\} \cup i(E) \times I & \overset{h}{\longrightarrow} & E \\
{\scriptstyle in} \downarrow & {\scriptstyle G} & \downarrow {\scriptstyle \pi} \\
E \times B \times I & \overset{g}{\longrightarrow} & B
\end{array}
$$

The upper horizontal map $h$ is given on $E \times B \times \{0\}$ by projection to $E \times \{0\} \subset E \times I$ composed with $F$, while the composite of $i \times id$ and $h$ equals $F$.

We shall construct $G$ by the same inductive procedure as before. To start the induction, define $G$ on $E \times \{v\} \times I$ to be a lift in the diagram

$$
\begin{array}{ccc}
E \times \{v\} \times \{0\} \cup i(E_v) \times I & \overset{h_v}{\longrightarrow} & E_v \\
{\scriptstyle in} \downarrow & {\scriptstyle G} & \downarrow \\
E \times \{v\} \times I & \longrightarrow & \{v\},
\end{array}
$$

where $h_v$ is the restriction of $h$. This is always possible, since $i(E_v) = E_v \times \{v\}$ is a CW subcomplex of $E \times \{v\}$.

For the inductive step, consider $\Delta$ and $\Lambda$ as before, and suppose that $G \colon E \times \Lambda \times I \to E_\Lambda$ under $E \times \Lambda \times \{0\} \cup i(E_\Lambda) \times I$ and over $\Lambda$ has already been constructed. The desired extension $G \colon E \times \Delta \times I \to E_\Delta$ is a lift in the following diagram.

$$
\begin{array}{ccc}
E \times \Delta \times \{0\} \cup (i(E_\Delta) \cup E \times \Lambda) \times I & \longrightarrow & E_\Delta \\
{\scriptstyle in} \downarrow & & \downarrow {\scriptstyle \pi_\Delta} \\
E \times \Delta \times I & \longrightarrow & \Delta
\end{array}
$$

Here the upper horizontal map equals the restriction of $h$ on $E \times \Delta \times \{0\} \cup i(E_\Delta) \times I$, and it equals the previously defined $G$ on $E \times \Lambda \times I$. The lower horizontal map is the restriction of $g$. Once more, such a lift exists because $i(E_\Delta) \cup E \times \Lambda$ is a CW subcomplex of $E \times \Delta$ and $\pi_\Delta$ is a Serre fibration. By induction, it follows that $G$ can be extended to all of $E \times B \times I$.

Now $r \colon E \times B \to E$ given by $r(e, b) = G(e, b, 1)$ is the required retraction over $B$. $\square$

We will not need to know whether a simplex-wise Serre fibration over a singular base must always be a Serre fibration.

# Chapter Three

## The non-manifold part

In this chapter, let $\Delta^q$ denote the simplicial $q$-simplex, the simplicial set represented by $[q]$ in $\Delta$, and let $|\Delta^q|$ denote its geometric realization, the standard affine $q$-simplex.

### 3.1. Categories of simple maps

**Definition 3.1.1.** Let $\mathcal{C}$ be the category of finite simplicial sets $X$ and simplicial maps $f \colon X \to Y$. Let $\mathcal{D}$ be the full subcategory of $\mathcal{C}$ of finite non-singular simplicial sets, and write $i \colon \mathcal{D} \to \mathcal{C}$ for the inclusion functor. Let $\mathcal{E}$ be the category of compact polyhedra $K$ and PL maps $f \colon K \to L$, and write $r \colon \mathcal{D} \to \mathcal{E}$ for the polyhedral realization functor. We also write $i$ and $r$ for other variants of these functors, such as in diagram (3.1.5).

The letter $\mathcal{E}$ might allude to **Euclidean** polyhedra, the non-singular simplicial sets in $\mathcal{D}$ may have been **desingularized** (see Remark 2.2.12), and the relative category $\mathcal{C}(X)$ to be introduced in Definition 3.1.6 is a category of **cofibrations** under $X$. For mnemonic reasons we shall use the letters (c), (d) and (e) to enumerate these three contexts, even if we have not used the letters (a) and (b) before.

In more detail, the **polyhedral realization** of a finite non-singular simplicial set $X$ is its geometric realization $|X|$, which is compact, with the polyhedral structure for which the geometric realization $|\bar{x}| \colon |\Delta^q| \to |X|$ of the representing map of each simplex $x \in X_q$ is a PL map. See also Definition 3.4.1. For a singular (= not non-singular) simplicial set $X$, this prescription does not in general define a polyhedral structure on $|X|$.

**Definition 3.1.2.** By a **PL bundle** $\pi \colon E \to B$, we mean a PL map of compact polyhedra that admits a PL local trivialization.

(d) Let $\mathcal{D}_\bullet$ be the simplicial category that in simplicial degree $q$ consists of the finite non-singular simplicial sets over $\Delta^q$, i.e., the simplicial maps $\pi \colon Z \to \Delta^q$ whose polyhedral realization $|\pi| \colon |Z| \to |\Delta^q|$ is a PL bundle, and the simplicial maps $f \colon Z \to Z'$ over $\Delta^q$.

(e) Let $\mathcal{E}_\bullet$ be the simplicial category of PL bundles $\pi \colon E \to |\Delta^q|$, and PL bundle maps $f \colon E \to E'$ over $|\Delta^q|$, in each simplicial degree $q$. The simplicial structure is given by pullback of PL bundles along the base.

We do not define a simplicial category $\mathcal{C}_\bullet$, because it makes little sense to ask that $|\pi| \colon |Z| \to |\Delta^q|$ is a PL bundle when $|Z|$ is not a polyhedron.

**Definition 3.1.3.** A map of topological spaces is a Serre fibration if it has the homotopy lifting property for continuous maps from compact polyhedra. A map of simplicial sets is called a **Serre fibration** if its geometric realization is a Serre fibration of topological spaces. By a **PL Serre fibration** $\pi \colon E \to B$, we mean a PL map of compact polyhedra that is a Serre fibration in the topological category.

(c) Let $\widetilde{\mathcal{C}}_\bullet$ be the simplicial category with objects the Serre fibrations of finite simplicial sets $\pi \colon Z \to \Delta^q$, and with morphisms the simplicial maps $f \colon Z \to Z'$ over $\Delta^q$, in each simplicial degree $q$.

(d) Let $\widetilde{\mathcal{D}}_\bullet$ be the simplicial category of Serre fibrations of finite non-singular simplicial sets $\pi \colon Z \to \Delta^q$, and simplicial maps $f \colon Z \to Z'$ over $\Delta^q$, in each simplicial degree $q$.

(e) Let $\widetilde{\mathcal{E}}_\bullet$ be the simplicial category of PL Serre fibrations $\pi \colon E \to |\Delta^q|$, and PL maps $f \colon E \to E'$ over $|\Delta^q|$, in each simplicial degree $q$. The simplicial structure is given by pullback of PL Serre fibrations along the base.

There are canonical identifications $\mathcal{C} \cong \widetilde{\mathcal{C}}_0$, $\mathcal{D} \cong \mathcal{D}_0 = \widetilde{\mathcal{D}}_0$ and $\mathcal{E} \cong \mathcal{E}_0 = \widetilde{\mathcal{E}}_0$. The inclusion of 0-simplices, via degeneracies, induces functors $\tilde{n} \colon \mathcal{C} \to \widetilde{\mathcal{C}}_\bullet$, $n \colon \mathcal{D} \to \mathcal{D}_\bullet$, $\tilde{n} \colon \mathcal{D} \to \widetilde{\mathcal{D}}_\bullet$, $n \colon \mathcal{E} \to \mathcal{E}_\bullet$ and $\tilde{n} \colon \mathcal{E} \to \widetilde{\mathcal{E}}_\bullet$ ($n$ for null). For example, $n \colon \mathcal{E} \to \mathcal{E}_\bullet$ takes a compact polyhedron $K$ to the product PL bundle $pr \colon K \times |\Delta^q| \to |\Delta^q|$, and $\tilde{n} \colon \mathcal{E} \to \widetilde{\mathcal{E}}_\bullet$ takes $K$ to the product PL Serre fibration $pr \colon K \times |\Delta^q| \to |\Delta^q|$. There are full embeddings $v \colon \mathcal{D}_\bullet \to \widetilde{\mathcal{D}}_\bullet$ and $v \colon \mathcal{E}_\bullet \to \widetilde{\mathcal{E}}_\bullet$ that view PL bundles as PL Serre fibrations, and $\tilde{n} = vn$ in these two cases.

**Definition 3.1.4.** Let $s\mathcal{C} \subset \mathcal{C}$ and $s\mathcal{D} \subset \mathcal{D}$ be the subcategories with morphisms the simple maps $f \colon X \to Y$ of finite (non-singular) simplicial sets, and let $s\mathcal{E} \subset \mathcal{E}$ be the subcategory with morphisms the simple maps $f \colon K \to L$ of compact polyhedra.

Likewise, let $s\widetilde{\mathcal{C}}_\bullet \subset \widetilde{\mathcal{C}}_\bullet$, $s\mathcal{D}_\bullet \subset \mathcal{D}_\bullet$ and $s\widetilde{\mathcal{D}}_\bullet \subset \widetilde{\mathcal{D}}_\bullet$ be the subcategories with morphisms the simple maps $f \colon Z \to Z'$ of finite simplicial sets, and let $s\mathcal{E}_\bullet \subset \mathcal{E}_\bullet$ and $s\widetilde{\mathcal{E}}_\bullet \subset \widetilde{\mathcal{E}}_\bullet$ be the subcategories with morphisms the simple maps $f \colon E \to E'$ of compact polyhedra, always in each simplicial degree $q$.

Similarly, let $h\mathcal{C} \subset \mathcal{C}$, etc., be the subcategory with morphisms the weak homotopy equivalences $f \colon X \to Y$ of finite simplicial sets, and let $h\mathcal{E} \subset \mathcal{E}$, etc., be the subcategory with morphisms the homotopy equivalences $f \colon K \to L$ of compact polyhedra. Write $j \colon s\mathcal{C} \to h\mathcal{C}$, $j \colon s\widetilde{\mathcal{E}}_\bullet \to h\widetilde{\mathcal{E}}_\bullet$, etc., for the inclusion functors that view simple maps as (weak) homotopy equivalences.

By Proposition 2.1.3(a), simple maps of finite simplicial sets can be composed, and therefore form a category. We assemble the $s$-prefixed categories in the following commutative diagram.

(3.1.5)

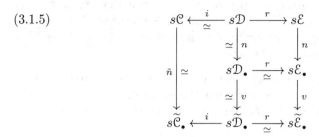

It maps by many variants of the functor $j$ to a similar diagram with $h$-prefixes. We shall show in Proposition 3.1.14 that $i$ in the top row is a homotopy equivalence, in Proposition 3.4.4 that $r$ in the middle and lower rows are homotopy equivalences, and in Proposition 3.5.1 that $n$ and $v$ in the middle column, as well as $\tilde{n}$ in the left hand column, are homotopy equivalences. It follows that $v$ in the right hand column and $i$ in the bottom row are homotopy equivalences. These results apply equally well in the $s$- and $h$-cases.

We do not know whether $r\colon s\mathcal{D} \to s\mathcal{E}$ and $n\colon s\mathcal{E} \to s\mathcal{E}_{\bullet}$ are homotopy equivalences. This is equivalent to the statement of [Ha75, Prop. 2.5], which remains unproven.

Our principal interest is in the corresponding diagram of homotopy fibers of $j$ at a finite non-singular simplicial set $X$, or at its polyhedral realization $|X|$. The strategy for studying these homotopy fibers is to use Quillen's Theorem B [Qu73, p. 97] in its form for right fibers $X/j$, or the simplicial version Theorem B' from [Wa82, §4] in the cases where we are dealing with simplicial categories. These theorems are recalled in Section 3.6 at the end of this chapter.

To accomplish this study we shall make use of approximations to these right fibers, with names like $s\mathcal{C}^h(X)$, that have better covariant functoriality properties than the right fibers themselves, but still retain some of the contravariant functoriality properties of the latter. We now introduce these approximations, which are relative forms of the categories introduced so far.

**Definition 3.1.6.** (c) To each finite simplicial set $X$, we associate a category $\mathcal{C}(X)$ of finite cofibrations under $X$. Its objects are the cofibrations $y\colon X \to Y$ of finite simplicial sets, and the morphisms from $y$ to $y'\colon X \to Y'$ are the maps $f\colon Y \to Y'$ of simplicial sets under $X$, i.e., the simplicial maps such that $fy = y'$. Let $s\mathcal{C}^h(X) \subset \mathcal{C}(X)$ be the subcategory where the objects $y\colon X \to Y$ are required to be weak homotopy equivalences, and the morphisms $f\colon Y \to Y'$ are required to be simple maps.

(d) Let $\mathcal{D}(X) \subset \mathcal{C}(X)$ be the full subcategory of cofibrations $y\colon X \to Y$ of finite non-singular simplicial sets. This is only of real interest when $X$ itself is non-singular, because otherwise $\mathcal{D}(X)$ is the empty category. Let $s\mathcal{D}^h(X) \subset s\mathcal{C}^h(X)$ be the full subcategory generated by the objects $y\colon X \to Y$ where $Y$ is non-singular.

(e) To each compact polyhedron $K$, we associate a category $\mathcal{E}(K)$ of compact polyhedra containing $K$. Its objects are the PL embeddings $\ell\colon K \to L$ of compact polyhedra, and the morphisms from $\ell$ to $\ell'\colon K \to L'$ are the PL maps $f\colon L \to L'$ that restrict to the identity on $K$, i.e., such that $f\ell = \ell'$. Let $s\mathcal{E}^h(K) \subset \mathcal{E}(K)$ be the subcategory where the PL embedding $\ell\colon K \to L$ is required to be a homotopy equivalence (for the objects), and $f\colon L \to L'$ is required to be a simple map (for the morphisms).

If desired, one may adjust these definitions to ask that the cofibration $y\colon X \to Y$ is an inclusion $X \subset Y$, and similarly that the embedding $\ell\colon K \to L$ is the inclusion of a subpolyhedron $K \subset L$. The resulting categories are canonically equivalent.

**Definition 3.1.7.** (c) To each finite simplicial set $X$, we associate a simplicial category $\widetilde{\mathcal{C}}_\bullet(X)$. Its objects in simplicial degree $q$ are commutative diagrams of finite simplicial sets

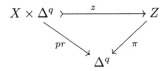

where $z$ is a cofibration and $\pi$ is a Serre fibration. The morphisms in simplicial degree $q$, to a similar object more briefly denoted

$$pr\colon X \times \Delta^q \xrightarrow{z'} Z' \xrightarrow{\pi'} \Delta^q\,,$$

are the simplicial maps $f\colon Z \to Z'$ over $\Delta^q$ that restrict to the identity on $X \times \Delta^q$, i.e., such that $\pi'f = \pi$ and $fz = z'$.

Let $s\widetilde{\mathcal{C}}_\bullet^h(X) \subset \widetilde{\mathcal{C}}_\bullet(X)$ be the simplicial subcategory where $z\colon X \times \Delta^q \to Z$ is required to be a weak homotopy equivalence (for the objects), and $f\colon Z \to Z'$ is required to be a simple map (for the morphisms).

(d) Let $\mathcal{D}_\bullet(X)$ (resp. $\widetilde{\mathcal{D}}_\bullet(X)$) be the full simplicial subcategory of $\widetilde{\mathcal{C}}_\bullet(X)$ with objects the commutative diagrams of finite non-singular simplicial sets

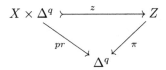

where $z$ is a cofibration, and $|\pi|\colon |Z| \to |\Delta^q|$ is a PL bundle relative to the product subbundle $pr\colon |X| \times |\Delta^q| \to |\Delta^q|$ (resp. a PL Serre fibration).

Let $s\mathcal{D}_\bullet^h(X) \subset \mathcal{D}_\bullet(X)$ and $s\widetilde{\mathcal{D}}_\bullet^h(X) \subset \widetilde{\mathcal{D}}_\bullet(X)$ be the simplicial subcategories with objects such that $z$ is a weak homotopy equivalence, and morphisms such that $f$ is a simple map, in each simplicial degree. As before, these constructions are only of real interest when $X$ is non-singular.

(e) To each compact polyhedron $K$, we associate two simplicial categories $\mathcal{E}_\bullet(K)$ and $\widetilde{\mathcal{E}}_\bullet(K)$. The objects in simplicial degree $q$ of $\mathcal{E}_\bullet(K)$ (resp. $\widetilde{\mathcal{E}}_\bullet(K)$) are the commutative diagrams of compact polyhedra and PL maps

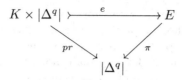

where $e$ is a PL embedding, and $\pi$ is a PL bundle relative to the product subbundle $pr: K \times |\Delta^q| \to |\Delta^q|$ (resp. a PL Serre fibration). The morphisms in simplicial degree $q$, to a similar object more briefly denoted

$$ pr: K \times |\Delta^q| \xrightarrow{e'} E' \xrightarrow{\pi'} |\Delta^q| , $$

are PL maps $f: E \to E'$ over $|\Delta^q|$ that restrict to the identity on $K \times |\Delta^q|$, i.e., such that $\pi'f = \pi$ and $fe = e'$.

Let $s\mathcal{E}_\bullet^h(K) \subset \mathcal{E}_\bullet(K)$ and $s\widetilde{\mathcal{E}}_\bullet^h(K) \subset \widetilde{\mathcal{E}}_\bullet(K)$ be the simplicial subcategories where for the objects the embedding $e: K \times |\Delta^q| \to E$ is required to be a homotopy equivalence, and for the morphisms the PL map $f: E \to E'$ is required to be a simple map.

Recall from Section 1.1 that for $\pi$ to be a PL bundle relative to a given product subbundle means that $\pi$ admits a PL local trivialization that agrees with the identity map on the embedded subbundle. In each case, a homotopy equivalence $e: K \times |\Delta^q| \to E$ is also a fiberwise homotopy equivalence over $|\Delta^q|$, and conversely, because $pr$ and $\pi$ are Serre fibrations. So the objects in simplicial degree $q$ of $s\mathcal{E}_\bullet^h(K)$ or $s\widetilde{\mathcal{E}}_\bullet^h(K)$ are really $q$-parameter families of objects of $s\mathcal{E}^h(K)$. Similar remarks apply in the (d)-case.

For each finite non-singular simplicial set $X$, these categories fit together in a commutative diagram

(3.1.8)

$$
\begin{array}{ccccc}
s\mathcal{C}^h(X) & \xleftarrow[\simeq]{i} & sD^h(X) & \xrightarrow{r} & s\mathcal{E}^h(|X|) \\
& & & & \\
\tilde{n} \downarrow \simeq & vn \atop \simeq & sD_\bullet^h(X) \xrightarrow{r} s\mathcal{E}_\bullet^h(|X|) & & \downarrow n \\
& & \downarrow v & & \downarrow v \\
s\widetilde{\mathcal{C}}_\bullet^h(X) & \xleftarrow{i} & s\widetilde{D}_\bullet^h(X) & \xrightarrow[\simeq]{r} & s\widetilde{\mathcal{E}}_\bullet^h(|X|)
\end{array}
$$

that maps to diagram (3.1.5) by functors that forget the structure map from $X$ or $|X|$. The composite functor to the $h$-prefixed version of (3.1.5) is then canonically homotopic to the constant functor to the object $X$ or $|X|$, by the natural transformation represented by the structure map to be forgotten. Hence there are well-defined maps

(3.1.9)                  $s\mathcal{C}^h(X) \to X/j \to \mathrm{hofib}(j\colon s\mathcal{C} \to h\mathcal{C}, X)$

from (the nerve of) the relative category at $X$, via (the nerve of) the right fiber of $j$ at $X$, to the homotopy fiber of $j$ at $X$, viewed as an object of $h\mathcal{C}$. There are analogous maps from each of the other categories in (3.1.8). In Proposition 3.3.1 and Lemma 3.3.3 we shall show that both maps in (3.1.9) are homotopy equivalences, for each of the categories $s\mathcal{C}^h(X)$, $s\mathcal{D}^h(X)$, $s\mathcal{E}^h(K)$ in the top row, and $s\widetilde{\mathcal{C}}^h_\bullet(X)$, $s\widetilde{\mathcal{D}}^h_\bullet(X)$ and $s\widetilde{\mathcal{E}}^h_\bullet(K)$ in bottom row.

    It follows from (3.1.5), its $h$-analog, and these homotopy fiber sequences, that the indicated functors $i$, $vn$, $r$ and $\tilde{n}$ are homotopy equivalences. This will be spelled out in the proofs of Theorems 1.2.5 and 1.2.6.

*Remark 3.1.10.* We have also written down a proof that $v\colon s\mathcal{D}^h_\bullet(X) \to s\widetilde{\mathcal{D}}^h_\bullet(X)$ is a homotopy equivalence. Thus each of the functors $v$ and $n$ in the middle column of (3.1.8), not just their composite, is a homotopy equivalence. Together with part of Proposition 3.5.1(d), this implies that the composite map from $s\mathcal{D}^h_\bullet(X)$ to the homotopy fiber of $j\colon s\mathcal{D}_\bullet \to h\mathcal{D}_\bullet$ at $X$ is a homotopy equivalence. However, the details of our argument are unpleasant, and so we omit this proof because we do not need to use the result anyway. We do not know whether the composite map from $s\mathcal{E}^h_\bullet(K)$ to the homotopy fiber of $j\colon s\mathcal{E}_\bullet \to h\mathcal{E}_\bullet$ at $K$ is a homotopy equivalence.

**Definition 3.1.11.** (c) The categories $\mathcal{C}(X)$ and $s\mathcal{C}^h(X)$ are covariantly functorial in $X$. A map $x\colon X \to X'$ of simplicial sets induces a (forward) functor

$$x_*\colon \mathcal{C}(X) \to \mathcal{C}(X')$$

that takes $y\colon X \to Y$ to the pushout $y'\colon X' \to X' \cup_X Y$. For any map $x'\colon X' \to X''$ there is a coherent natural isomorphism $(x'x)_* \cong x'_* x_*$. (Strictly speaking, for $X \mapsto \mathcal{C}(X)$ to be a functor, this isomorphism should be the identity. This can be arranged by a careful definition of the pushout.) By the gluing lemmas for simple maps and for weak homotopy equivalences of simplicial sets, the functor $x_*$ restricts to a functor

$$x_*\colon s\mathcal{C}^h(X) \to s\mathcal{C}^h(X').$$

Corresponding gluing lemmas are not available to define a functor $x_*\colon X/j \to X'/j$ of right fibers for $j\colon s\mathcal{C} \to h\mathcal{C}$.

    There is also a functor

$$x_*\colon s\widetilde{\mathcal{C}}^h_\bullet(X) \to s\widetilde{\mathcal{C}}^h_\bullet(X')$$

that takes an object $pr\colon \Delta^q \times X \xrightarrow{z} Z \xrightarrow{\pi} \Delta^q$ to the object $pr\colon \Delta^q \times X' \xrightarrow{z'} Z' \xrightarrow{\pi'} \Delta^q$ with total space

$$Z' = (X' \times \Delta^q) \cup_{(X \times \Delta^q)} Z.$$

The induced map of pushouts, $\pi' : Z' \to \Delta^q$, remains a Serre fibration by the fiber gluing lemma for Serre fibrations, Lemma 2.7.10.

(d) The categories $\mathcal{D}(X)$, $s\mathcal{D}^h(X)$, $s\mathcal{D}^h_\bullet(X)$ and $s\widetilde{\mathcal{D}}^h_\bullet(X)$ are also covariantly functorial in $X$, but only along cofibrations. A cofibration $x : X \to X'$ of simplicial sets induces a functor

$$x_* : \mathcal{D}(X) \to \mathcal{D}(X')$$

that takes $y : X \to Y$ to the pushout $y' : X' \to X' \cup_X Y$. We need to assume that $x$ and $y$ are cofibrations to be sure that $X' \cup_X Y$ remains non-singular. By the same gluing lemmas as before, there is a restricted functor

$$x_* : s\mathcal{D}^h(X) \to s\mathcal{D}^h(X') .$$

By the assumption of PL local triviality of $\pi$ relative to $pr : X \times \Delta^q \to \Delta^q$, after polyhedral realization, there is a functor $x_* : s\mathcal{D}^h_\bullet(X) \to s\mathcal{D}^h_\bullet(X')$ that takes an object $pr : \Delta^q \times X \xrightarrow{z} Z \xrightarrow{\pi} \Delta^q$ to the object $pr : \Delta^q \times X' \xrightarrow{z'} Z' \xrightarrow{\pi'} \Delta^q$ with total space

$$Z' = (X' \times \Delta^q) \cup_{(X \times \Delta^q)} Z .$$

The local trivializations can be glued to make $\pi'$ PL locally trivial relative to the trivial subbundle, after polyhedral realization, as required. There is also a functor

$$x_* : s\widetilde{\mathcal{D}}^h_\bullet(X) \to s\widetilde{\mathcal{D}}^h_\bullet(X') ,$$

by the fiber gluing lemma for Serre fibrations.

(e) For a compact polyhedron $K$, the categories $\mathcal{E}(K)$, $s\mathcal{E}^h(K)$, $s\mathcal{E}^h_\bullet(K)$ and $s\widetilde{\mathcal{E}}^h_\bullet(K)$ are also covariantly functorial in $K$, but only along PL embeddings. A PL embedding $k : K \to K'$ induces the basic functor

$$k_* : \mathcal{E}(K) \to \mathcal{E}(K')$$

that takes $\ell : K \to L$ to the union $\ell' : K' \to K' \cup_K L$. As usual, we need that both $k$ and $\ell$ are embeddings to have a canonical PL structure on the pushout. By the same arguments as in the $\mathcal{D}$-case, there are variants for $s\mathcal{E}^h(K)$, $s\mathcal{E}^h_\bullet(K)$ and

$$k_* : s\widetilde{\mathcal{E}}^h_\bullet(K) \to s\widetilde{\mathcal{E}}^h_\bullet(K') .$$

**Definition 3.1.12.** We can extend the definition of $\mathcal{C}(X)$, $\mathcal{D}(X)$, etc., to infinite simplicial sets $X$. By a **finite cofibration** $y : X \to Y$ we mean a cofibration such that $Y$ only contains finitely many non-degenerate simplices that are not in the image of $X$.

For a general simplicial set $X$, we let $\mathcal{C}(X)$ be the category of finite cofibrations $y : X \to Y$, and simplicial maps $f : Y \to Y'$ under $X$. Since $Y$ and $Y'$ will no longer be finite simplicial sets, we must specify what we mean by a simple map $f : Y \to Y'$. There is a canonical identification

$$\operatorname*{colim}_\alpha \mathcal{C}(X_\alpha) \cong \mathcal{C}(X) ,$$

where $X_\alpha$ ranges over the set of finite simplicial subsets of $X$, partially ordered by inclusion. We then interpret $s\mathcal{C}^h(X)$ as the subcategory

$$\operatorname*{colim}_{\alpha} s\mathcal{C}^h(X_\alpha).$$

An object of $s\mathcal{C}^h(X)$ is thus a finite cofibration and weak homotopy equivalence $y\colon X \to Y$, and a morphism $f$ from $y$ to $y'\colon X \to Y'$ restricts to the identity on $X$ and has contractible point inverses after geometric realization. From the interpretation as a colimit, it is clear that such morphisms can be composed, so that $s\mathcal{C}^h(X)$ is indeed a category.

The right fiber $X/j$, for $j\colon s\mathcal{C} \to h\mathcal{C}$, say, is contravariantly functorial in the object $X$ of $h\mathcal{C}$. So each weak homotopy equivalence $x\colon X \to X'$ induces a functor $x^*\colon X'/j \to X/j$, known as a **transition map**. The relative categorical constructions are also contravariantly functorial in $X$, but only in a slightly more limited sense.

**Definition 3.1.13.** (c) A finite cofibration $x\colon X \to X'$ induces a backward functor

$$x^*\colon \mathcal{C}(X') \to \mathcal{C}(X)$$

that takes a finite cofibration $y'\colon X' \to Y$ to the composite $y'x\colon X \to Y$. The relation $(x'x)^* = x^*x'^*$ holds for any finite cofibration $x'\colon X' \to X''$. If $x$ is also a weak homotopy equivalence, then there is also a restricted functor

$$x^*\colon s\mathcal{C}^h(X') \to s\mathcal{C}^h(X).$$

Similarly for the constructions $\widetilde{\mathcal{C}}_\bullet(-)$ and $s\widetilde{\mathcal{C}}^h_\bullet(-)$.

(d) If $X$ and $X'$ are finite non-singular simplicial sets, there are backward functors $x^*\colon \mathcal{D}(X') \to \mathcal{D}(X)$ (when $x$ is a cofibration) and $x^*\colon s\mathcal{D}^h(X') \to s\mathcal{D}^h(X)$ (when $x$ is also a weak homotopy equivalence), and likewise for the constructions $\mathcal{D}_\bullet(-)$, $\widetilde{\mathcal{D}}_\bullet(-)$, $s\mathcal{D}^h_\bullet(-)$ and $s\widetilde{\mathcal{D}}^h_\bullet(-)$.

(e) A PL embedding $k\colon K \to K'$ of compact polyhedra induces a backward functor

$$k^*\colon \mathcal{E}(K') \to \mathcal{E}(K)$$

that takes a PL embedding $\ell'\colon K' \to L$ to the composite $\ell'k\colon K \to L$. A PL embedding and homotopy equivalence $k\colon K \to K'$ induces a restricted functor

$$k^*\colon s\mathcal{E}^h(K') \to s\mathcal{E}^h(K).$$

There are similar backward functors $k^*$ in the cases $\mathcal{E}_\bullet(-)$, $\widetilde{\mathcal{E}}_\bullet(-)$, $s\mathcal{E}^h_\bullet(-)$ and $s\widetilde{\mathcal{E}}^h_\bullet(-)$.

We can now prove the part of the stable parametrized $h$-cobordism theorem that compares general and non-singular simplicial sets, i.e., Theorem 1.2.5.

**Proposition 3.1.14.** *Let $X$ be a finite non-singular simplicial set. The full inclusion functors induce homotopy equivalences* $i\colon s\mathcal{D} \to s\mathcal{C}$, $i\colon h\mathcal{D} \to h\mathcal{C}$ *and*

$$i\colon s\mathcal{D}^h(X) \to s\mathcal{C}^h(X).$$

We offer two proofs, one using Quillen's Theorem A, and one using the improvement functor $I = B \circ Sd$ from Section 2.5. The former proof is essentially the same as that given in [St86, Thm. 3.1], except that we include more details. The latter proof depends on the more difficult parts of Sections 2.3 and 2.5, but has the advantage that it provides a functorial homotopy inverse to $i$.

*First proof.* We first prove that $i\colon s\mathcal{D} \to s\mathcal{C}$ is a homotopy equivalence. Using Quillen's Theorem A, it suffices to show that the left fiber $i/Y$ is contractible, for each finite simplicial set $Y$. By Proposition 2.5.1(a), there exists a finite non-singular simplicial set $Z$ and a simple map $f\colon Z \to Y$. Let $f_*\colon i/Z \to i/Y$ be given by composition with $f$, and define $f^*\colon i/Y \to i/Z$ to be the functor taking a simple map $W \to Y$ (from a finite non-singular $W$) to the pullback $Z \times_Y W \to Z$.

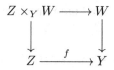

The fiber product $Z \times_Y W$ is finite and non-singular, being contained in the product $Z \times W$, and the map to $Z$ is simple by the pullback property of simple maps. The map $Z \times_Y W \to W$ is also simple, for the same reason, and defines a natural transformation $f_* f^* \to id$ of endofunctors of $i/Y$. Hence $i/Y$ is a retract, up to homotopy, of $i/Z$. The latter category has a terminal object, namely $id_Z$, hence is contractible. Thus also $i/Y$ is contractible.

The proof that $i\colon h\mathcal{D} \to h\mathcal{C}$ is a homotopy equivalence is similar. To see that the pullback $Z \times_Y W \to Z$ of a weak homotopy equivalence $W \to Y$ is a weak homotopy equivalence, note that $f$ and its pullback $Z \times_Y W \to W$ are simple, and use the 2-out-of-3 property of weak homotopy equivalences.

To prove that $i\colon s\mathcal{D}^h(X) \to s\mathcal{C}^h(X)$ is a homotopy equivalence, it suffices to show that $i/(X \to Y)$ is contractible, for any object $X \to Y$ of $s\mathcal{C}^h(X)$. By Proposition 2.5.1(b) there exists a second such object $X \to Z$, with $Z$ non-singular, and a simple map $f\colon Z \to Y$, relative to $X$. By the same constructions as above, there are functors $f_*\colon i/(X \to Z) \to i/(X \to Y)$ and $f^*\colon i/(X \to Y) \to i/(X \to Z)$, and a natural transformation $f_* f^* \to id$. The required finite cofibration and weak homotopy equivalence $X \to Z \times_Y W$ is obtained as the pullback of the maps $X \to Z$ and $X \to W$ (over $X \to Y$), of the same kind. As before, $i/(X \to Z)$ has a terminal object and is contractible, so also its homotopy retract $i/(X \to Y)$ is contractible. $\square$

Alternatively, we could have deduced the claim for $s\mathcal{D}^h(X) \to s\mathcal{C}^h(X)$ from the other two claims and the map of horizontal fiber sequences

(3.1.15)
$$
\begin{array}{ccc}
s\mathcal{D}^h(X) \longrightarrow s\mathcal{D} \xrightarrow{\ j\ } h\mathcal{D} \\
\downarrow{\scriptstyle i} \qquad \downarrow{\scriptstyle i} \qquad \downarrow{\scriptstyle i} \\
s\mathcal{C}^h(X) \longrightarrow s\mathcal{C} \xrightarrow{\ j\ } h\mathcal{C}
\end{array}
$$

from Proposition 3.3.1(c) and (d). Conversely, Propositions 3.1.14 and 3.3.1(c), provide a proof of the $s\mathcal{D}^h(X)$-part of Proposition 3.3.1(d), which does not depend on the somewhat complicated non-singular case (d) of Lemma 3.2.9. We do not know if the $s\widetilde{\mathcal{D}}^h_\bullet(X)$-part of Proposition 3.3.1(d) can be deduced in this way.

*Second proof.* The improvement functor $I$ of Theorem 2.5.2 defines a homotopy inverse $I\colon s\mathcal{C} \to s\mathcal{D}$ to the full inclusion $i\colon s\mathcal{D} \to s\mathcal{C}$, because the natural simple map $s_X\colon I(X) \to X$ defines a natural transformation from each of the composites $iI$ and $Ii$ to the respective identities. Similarly, $I$ defines a homotopy inverse $I\colon h\mathcal{C} \to h\mathcal{D}$ to $i\colon h\mathcal{D} \to h\mathcal{C}$.

For the relative case, we define a homotopy inverse functor $s\mathcal{C}^h(X) \to s\mathcal{D}^h(X)$ to $i\colon s\mathcal{D}^h(X) \to s\mathcal{C}^h(X)$ by

$$
(X \to Y) \longmapsto \left(X \to M(s_X) \cup_{I(X)} I(Y)\right).
$$

Here $M(s_X)$ is the reduced mapping cylinder of the simple map $s_X\colon I(X) \to X$, which contains $X$ and $I(X)$ via the back and front inclusions, respectively. The pushout $M(s_X) \cup_{I(X)} I(Y)$ is finite and non-singular by Lemma 2.4.11(c) and Theorem 2.5.2, and $X$ maps to it by a cofibration and weak homotopy equivalence, so it really does define an object of $s\mathcal{D}^h(X)$. The cylinder projection $pr\colon M(s_X) \to X$ is simple by Lemma 2.4.8, so by the gluing lemma its pushout with the simple maps $s_X\colon I(X) \to X$ and $s_Y\colon I(Y) \to Y$ is a simple map

$$
M(s_X) \cup_{I(X)} I(Y) \to X \cup_X Y = Y.
$$

It defines a natural transformation from the composite functor $s\mathcal{C}^h(X) \to s\mathcal{C}^h(X)$ to the identity, and restricts to a similar natural transformation from the composite functor on $s\mathcal{D}^h(X)$, because the latter is a full subcategory. Thus the full inclusion $i\colon s\mathcal{D}^h(X) \to s\mathcal{C}^h(X)$ is a homotopy equivalence. $\square$

## 3.2. FILLING HORNS

**Definition 3.2.1.** Following [Ka57, §10], we define the *$i$-th horn* $\Lambda^n_i$, for $n \geq 1$ and $0 \leq i \leq n$, to be the simplicial subset of $\Delta^n$ generated by all the proper faces that contain the $i$-th vertex. It is the cone in $\Delta^n$ with vertex $i$ and base the boundary of the $i$-th face. The inclusion $\Lambda^n_i \subset \Delta^n$ is a weak homotopy equivalence.

(c) A simplicial map $h\colon \Lambda_i^n \to X$ will be called a **horn in** $X$. If $X' = \Delta^n \cup_{\Lambda_i^n} X$ is the pushout of maps

$$\Delta^n \supset \Lambda_i^n \xrightarrow{h} X,$$

then we say that $X'$ is obtained from $X$ by **filling a horn**. We say that a finite cofibration $x\colon X \to X'$ can be obtained by filling finitely many horns if it factors as the composite of finitely many such inclusions $X \subset \Delta^n \cup_{\Lambda_i^n} X$, up to isomorphism.

(d) A cofibration $h\colon \Lambda_i^n \to X$ will be called an **embedded horn in** $X$. If $X$ is a non-singular simplicial set and $h$ is an embedded horn in $X$, then $X' = \Delta^n \cup_{\Lambda_i^n} X$ will also be non-singular, and we say that $X'$ is obtained from $X$ by filling an embedded horn, or that $X \subset X'$ is an **elementary simplicial expansion**. A finite composite of elementary simplicial expansions is called a **simplicial expansion**. We also say that such a map is obtained by filling finitely many embedded horns.

(e) If $K$ is a compact polyhedron, and $h\colon |\Lambda_i^n| \to K$ is a PL embedding, then $K' = |\Delta^n| \cup_{|\Lambda_i^n|} K$ is also a compact polyhedron, and we say that $K \subset K'$ is an **elementary expansion**. If there is a finite chain of elementary expansions from $K$ to $K'$, we say that $K \subset K'$ is an **expansion**.

An early study of elementary expansions, elementary collapses (their formal inverses) and finite compositions of these, was made by I. Johansson [Jo32]. A deeper study, founding the subject of simple homotopy theory, was made by J. H. C. Whitehead [Wh39].

**Proposition 3.2.2.** *(c) If $X$ is a simplicial set and $x\colon X \to X'$ is obtained by filling finitely many horns, then the functors*

$$x_*\colon s\mathcal{C}^h(X) \to s\mathcal{C}^h(X') \qquad and \qquad x^*\colon s\mathcal{C}^h(X') \to s\mathcal{C}^h(X)$$

*are mutually inverse homotopy equivalences, and similarly for the construction $s\widetilde{\mathcal{C}}_\bullet^h(-)$.*

*(d) If $X$ is a non-singular simplicial set and $x\colon X \to X'$ is a simplicial expansion, then the functors $x_*\colon s\mathcal{D}^h(X) \to s\mathcal{D}^h(X')$ and $x^*\colon s\mathcal{D}^h(X') \to s\mathcal{D}^h(X)$ are mutually inverse homotopy equivalences, and similarly for the constructions $s\mathcal{D}_\bullet^h(-)$ and $s\widetilde{\mathcal{D}}_\bullet^h(-)$.*

*(e) If $K$ is a compact polyhedron and $k\colon K \to K'$ is an expansion, then the functors*

$$k_*\colon s\widetilde{\mathcal{E}}_\bullet^h(K) \to s\widetilde{\mathcal{E}}_\bullet^h(K') \qquad and \qquad k^*\colon s\widetilde{\mathcal{E}}_\bullet^h(K') \to s\widetilde{\mathcal{E}}_\bullet^h(K)$$

*are mutually inverse homotopy equivalences, and similarly for the constructions $s\mathcal{E}^h(-)$ and $s\mathcal{E}_\bullet^h(-)$.*

The proof will be given below, after Lemmas 3.2.6 through 3.2.8.

**Proposition 3.2.3.** *(c) If $x\colon X \to X'$ is a finite cofibration of simplicial sets, and a weak homotopy equivalence, then the functors*

$$x_*\colon s\mathcal{C}^h(X) \to s\mathcal{C}^h(X') \qquad and \qquad x^*\colon s\mathcal{C}^h(X') \to s\mathcal{C}^h(X)$$

*are both homotopy equivalences, and similarly for the construction $s\widetilde{\mathcal{C}}_\bullet^h(-)$.*

*(d) If $x\colon X \to X'$ is a finite cofibration of non-singular simplicial sets, and a weak homotopy equivalence, then the functors $x_*\colon s\mathcal{D}^h(X) \to s\mathcal{D}^h(X')$ and $x^*\colon s\mathcal{D}^h(X') \to s\mathcal{D}^h(X)$ are both homotopy equivalences, and similarly for the constructions $s\mathcal{D}_\bullet^h(-)$ and $s\widetilde{\mathcal{D}}_\bullet^h(-)$.*

*(e) If $k\colon K \to K'$ is a PL embedding of compact polyhedra, and a homotopy equivalence, then the functors*

$$k_*\colon s\widetilde{\mathcal{E}}_\bullet^h(K) \to s\widetilde{\mathcal{E}}_\bullet^h(K') \qquad and \qquad k^*\colon s\widetilde{\mathcal{E}}_\bullet^h(K') \to s\widetilde{\mathcal{E}}_\bullet^h(K)$$

*are both homotopy equivalences, and similarly for the constructions $s\mathcal{E}^h(-)$ and $s\mathcal{E}_\bullet^h(-)$.*

The proof is given below, after Lemma 3.2.9. In this generality, the homotopy equivalences $x_*$ and $x^*$ are **not** usually mutually inverse, as the proof of the following corollary makes clear.

There is a categorical sum pairing $s\mathcal{C}^h(X) \times s\mathcal{C}^h(X) \to s\mathcal{C}^h(X)$, which takes two finite cofibrations $y\colon X \to Y$ and $y'\colon X \to Y'$ to their pushout $y \cup y'\colon X \to Y \cup_X Y'$, and which induces a commutative monoid structure on $\pi_0 s\mathcal{C}^h(X)$. The following result was used in the proof of [Wa85, Prop. 3.1.1].

**Corollary 3.2.4.** *The commutative monoid of path components $\pi_0 s\mathcal{C}^h(X)$ is a group.*

*Proof.* For each object $y\colon X \to Y$ in $s\mathcal{C}^h(X)$ the endofunctor $y^* y_*$ is represented by pushout with $y$, and is a homotopy equivalence by Proposition 3.2.3. Thus the monoid sum in $\pi_0 s\mathcal{C}^h(X)$ with the class of $y$ is an isomorphism. In particular, the class of $y$ has an inverse in $\pi_0 s\mathcal{C}^h(X)$. $\square$

For connected $X$ with fundamental group $\pi$, the group in question is the Whitehead group $\mathrm{Wh}_1(\pi) = K_1(\mathbb{Z}[\pi])/(\pm\pi)$. This follows from the homotopy fiber sequence (1.3.3) and the isomorphism $\pi_1 A(X) \cong K_1(\mathbb{Z}[\pi])$.

The homotopy invariance of $s\mathcal{C}^h(-)$ can be extended as follows.

**Proposition 3.2.5.** *If $x\colon X \to X'$ is a weak homotopy equivalence of (arbitrary) simplicial sets, then the forward functor $x_*\colon s\mathcal{C}^h(X) \to s\mathcal{C}^h(X')$ is a homotopy equivalence.*

The proof will be given at the end of the section, after the long, three-part proof of Lemma 3.2.9. We now turn to the postponed proofs.

**Lemma 3.2.6.** *For each $n \geq 1$ and $0 \leq i \leq n$ there exists a finite non-singular simplicial set $A$ and a commutative diagram*

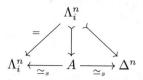

*in which the two maps originating from $A$ are simple, the map $\Lambda_i^n \to \Lambda_i^n$ is the identity and the map $\Lambda_i^n \to \Delta^n$ is the standard inclusion.*

*Proof.* We can define a cone $v\Delta^n$ with vertex $v$ and base $\Delta^n$ as the nerve of the disjoint union $[n] \sqcup \{v\}$, with a total ordering extending that of $[n] = \{0 < 1 < \cdots < n\}$. The simplicial set $A = v\Lambda_i^n$ is then defined as the subcone with vertex $v$ and base $\Lambda_i^n$.

To explain how the ordering relates $v$ to the elements of $[n]$, we make a case distinction. For $0 \leq i < n$ we declare that $i < v < i+1$, let $A \to \Lambda_i^n$ take $v$ to $i$, and let $A \to \Delta^n$ take $v$ to $i+1$. For $i = n$ we stipulate that $n-1 < v < n$, let $A \to \Lambda_i^n$ take $v$ to $n$, and let $A \to \Delta^n$ take $v$ to $n-1$. The point inverses of the geometric realizations of these maps are either single points or closed intervals, so the maps originating from $A$ are indeed simple. $\square$

**Lemma 3.2.7.** *(c) If $x\colon X \to X'$ is a map of simplicial sets given by filling a horn, then $x^* x_*$ is homotopic to the identity on $s\mathcal{C}^h(X)$, and similarly for $s\widetilde{\mathcal{C}}_\bullet^h(X)$.*

*(d) If $x\colon X \to X'$ is a map of non-singular simplicial sets given by filling an embedded horn, then $x^* x_*$ is homotopic to the identity on $s\mathcal{D}^h(X)$, and similarly for $s\mathcal{D}_\bullet^h(X)$ and $s\widetilde{\mathcal{D}}_\bullet^h(X)$.*

*(e) If $k\colon K \to K'$ is a map of polyhedra given by an elementary expansion, then $k^* k_*$ is homotopic to the identity on $s\widetilde{\mathcal{E}}_\bullet^h(K)$, and similarly for $s\mathcal{E}^h(K)$ and $s\mathcal{E}_\bullet^h(K)$.*

*Proof.* (c) The endofunctor $x^* x_*$ on $s\mathcal{C}^h(X)$ is given by pushout with the inclusion $\Lambda_i^n \to \Delta^n$. Let $\Phi$ be the endofunctor on the same category given by pushout with the map $\Lambda_i^n \to A$ of Lemma 3.2.6. By the gluing lemma for simple maps, the two simple maps $A \to \Lambda_i^n$ and $A \to \Delta^n$ of that lemma induce natural transformations $\Phi \to id$ and $\Phi \to x^* x_*$, respectively. These combine to give the required homotopy. The same argument applies for endofunctors of the simplicial category $s\widetilde{\mathcal{C}}_\bullet^h(X)$.

(d) Since $A$ is non-singular, the same proof applies in the non-singular case.

(e) The proof in the polyhedral case is similar, using the polyhedral realization of the diagram in Lemma 3.2.6. $\square$

**Lemma 3.2.8.** *For each $n \geq 1$ and $0 \leq i \leq n$ there is a finite non-singular*

*simplicial set B and a diagram*

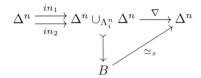

in which $\Delta^n \cup_{\Lambda_i^n} \Delta^n \to B$ is a simplicial expansion, $B \to \Delta^n$ is a simple map, the triangle commutes and the fold map $\nabla$ restricts to the identity on each copy of $\Delta^n$.

*Proof.* We take $B$ to be a cone with vertex $v$ and base $\Delta^n \cup_{\Lambda_i^n} \Delta^n$, obtained by gluing together two copies of the cone $v\Delta^n$ with vertex $v$ and base $\Delta^n$ along the subcone $A = v\Lambda_i^n$ from Lemma 3.2.6, with vertex $v$ and base $\Lambda_i^n$. Then $\Delta^n \cup_{\Lambda_i^n} \Delta^n \to B$ can be obtained by filling finitely many embedded horns.

In each of the two cones $v\Delta^n$ to be glued together, the ordering of the vertex $v$ with respect to the vertices $[n]$ of $\Delta^n$ is given by the same case distinction as in the cited lemma. The map $B \to \Delta^n$ is then given by gluing together two copies of a simple map $v\Delta^n \to \Delta^n$ along the simple map $A = v\Lambda_i^n \to \Delta^n$ from the same lemma, hence is also a simple map.    □

*Proof of Proposition 3.2.2.* (c) By induction, it suffices to consider the case of filling a single horn $x \colon X \to X' = \Delta^n \cup_{\Lambda_i^n} X$. By Lemma 3.2.7, we have left to show that $x_* x^*$ is homotopic to the identity on $s\mathcal{C}^h(X')$. Let

$$X'' = \Delta^n \cup_{\Lambda_i^n} X' = (\Delta^n \cup_{\Lambda_i^n} \Delta^n) \cup_{\Lambda_i^n} X$$

and let $x_1, x_2 \colon X' \to X''$ denote the two finite cofibrations that are induced from the two standard inclusions $in_1, in_2 \colon \Delta^n \to \Delta^n \cup_{\Lambda_i^n} \Delta^n$ by pushout along $\Lambda_i^n \to X$. Then $x_* x^* = x_2^* x_{1*}$. Let

$$X''' = B \cup_{(\Delta^n \cup_{\Lambda_i^n} \Delta^n)} X''$$

with $B$ as in Lemma 3.2.8, and let $x_3 \colon X'' \to X'''$ denote the inclusion. By the same lemma, $x_3$ can be obtained by filling finitely many embedded horns, so by Lemma 3.2.7 the composite $x_3^* x_{3*}$ (i.e., pushout with $x_3$) is homotopic to the identity on $s\mathcal{C}^h(X'')$. Consequently, $x_2^* x_{1*}$ is homotopic to $x_2^* x_3^* x_{3*} x_{1*} = (x_3 x_2)^* (x_3 x_1)_*$.

The latter composite may be identified with the endofunctor $\Phi$ of $s\mathcal{C}^h(X')$ given by pushout with $x_3 x_1 \colon X' \to X'''$ followed by switching the embedding of $X'$ to $x_3 x_2$, or what is the same, pushout with

$$\Delta^n \xrightarrow{in_1} \Delta^n \cup_{\Lambda_i^n} \Delta^n \to B$$

and switching the embedding of $\Delta^n$ to $in_2$. Therefore the simple map $B \to \Delta^n$ induces a natural transformation from $\Phi$ to the identity, giving the desired homotopy $(x_3 x_2)^* (x_3 x_1)_* \simeq id$.

The same argument applies for the construction $s\widetilde{\mathcal{C}}^h_\bullet(-)$.

(d) Since $B$ is non-singular, the same proof applies in the non-singular case.

(e) The proof in the polyhedral case is similar, using the polyhedral realization of the diagram in Lemma 3.2.8. □

To pass from maps obtained by filling finitely many horns, to general finite cofibrations that are weak homotopy equivalences, we use the following lemma. The three cases of its proof rely on rather different background material, so we first deduce Proposition 3.2.3 from it, and then recall the various concepts needed for the proof.

**Lemma 3.2.9.** *(c) Let $y\colon X \to Y$ be a finite cofibration and a weak homotopy equivalence of simplicial sets. Then there exists another finite cofibration (which is necessarily a weak homotopy equivalence) $z\colon Y \to Z$, such that $zy\colon X \to Z$ can be obtained by filling finitely many horns.*

*(d) Let $y\colon X \to Y$ be a finite cofibration and a weak homotopy equivalence of non-singular simplicial sets. Then there exists another finite cofibration of non-singular simplicial sets $z\colon Y \to Z$, such that $zy\colon X \to Z$ is a simplicial expansion.*

*(e) Let $\ell\colon K \to L$ be a PL embedding and homotopy equivalence of compact polyhedra. Then there exists another PL embedding $m\colon L \to M$, such that $m\ell\colon K \to M$ is an expansion.*

*Proof of Proposition 3.2.3.* We begin with the general simplicial case. Applying Lemma 3.2.9(c) twice, first to $x\colon X \to X'$ yielding $x'\colon X' \to X''$, and then to $x'\colon X' \to X''$ yielding $x''\colon X'' \to X'''$, we obtain a commutative diagram

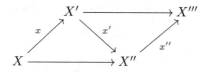

where the two horizontal arrows are inclusions obtained by filling finitely many horns. Applying Proposition 3.2.2 to these maps, we deduce that the forward functors $(x'x)_* \cong x'_*x_*$ and $(x''x')_* \cong x''_*x'_*$ are homotopy equivalences. Hence $x'_*$ has a right and a left homotopy inverse, and must therefore be a homotopy equivalence. Using again that $(x'x)_*$ is a homotopy equivalence, we conclude that $x_*$ is one, too. The dual argument applies for the backward functors.

The proofs in the non-singular and polyhedral cases are very similar, using Lemma 3.2.9(d) and (e), respectively. □

For the proof of the general simplicial case (c) of Lemma 3.2.9, we will use a particular fibrant replacement functor for simplicial sets.

A simplicial set is **fibrant** (or a **Kan complex**, or satisfies the **extension property** [Ka57, (1.1)]), if every horn can be filled within $X$, i.e., if every map $h\colon \Lambda^n_i \to X$ can be extended over the inclusion $\Lambda^n_i \to \Delta^n$. One way to enlarge a given simplicial set to a fibrant one is by repeatedly filling all horns

[GZ67, IV.3.2]. The following notation is meant to correspond to that for Kan's extension functor $Ex$, which was defined in [Ka57, §4], and its infinite iteration $Ex^\infty$. The left adjoint of the extension functor $Ex$ is the normal subdivision functor $Sd$ studied in Chapter 2.

**Definition 3.2.10.** For each simplicial set $X$ form $Fx(X)$ by **filling all horns**, i.e., by forming the pushout

$$
\begin{array}{ccc}
\coprod_{n,i,h} \Lambda_i^n & \longrightarrow & \coprod_{n,i,h} \Delta^n \\
\downarrow & & \downarrow \\
X & \xrightarrow{\ f_X\ } & Fx(X)
\end{array}
$$

where $n \geq 1$, $0 \leq i \leq n$ and $h$ ranges over all simplicial maps $h\colon \Lambda_i^n \to X$. This construction is functorial in $X$, because for any map $f\colon X \to Y$ and horn $h$ in $X$ to be filled in $Fx(X)$, the composite $fh$ is a horn in $Y$ to be filled in $Fx(Y)$. By the gluing lemma for weak homotopy equivalences, the cofibration $f_X\colon X \to Fx(X)$ is a weak homotopy equivalence, because each inclusion $\Lambda_i^n \to \Delta^n$ is one.

Let $Fx^0(X) = X$, $Fx^{r+1}(X) = Fx(Fx^r(X))$ for $r \geq 0$ and define

$$
Fx^\infty(X) = \operatorname*{colim}_r Fx^r(X)
$$

as the colimit of the diagram

$$
X \xrightarrow{\ f_X\ } Fx(X) \xrightarrow{\ f_{Fx(X)}\ } \cdots \to Fx^r(X) \xrightarrow{\ f_{Fx^r(X)}\ } Fx^{r+1}(X) \to \cdots .
$$

There is then a natural cofibration and weak homotopy equivalence

$$
f_X^\infty\colon X \to Fx^\infty(X),
$$

and $Fx^\infty(X)$ is fibrant. For any horn $h\colon \Lambda_i^n \to Fx^\infty(X)$ factors through $Fx^r(X)$ for some finite $r$, so $h$ can be filled in $Fx^{r+1}(X)$, and therefore also in $Fx^\infty(X)$. In other words, $Fx^\infty$ is a fibrant replacement functor.

*Proof of Lemma 3.2.9(c).* $Fx^\infty(X)$ is a fibrant simplicial set, so the map $f_X^\infty$ extends over the cofibration and weak homotopy equivalence $y\colon X \to Y$, see [GJ99, Thm. I.11.3]. Since $y$ is a finite cofibration, the resulting image of $Y$ in $Fx^\infty(X)$ is contained in a simplicial subset $W$ of $Fx^\infty(X)$ that is generated by $X$ and finitely many other simplices, and which can be obtained from $X$ by filling finitely many horns.

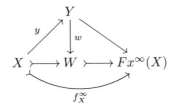

To turn $w\colon Y \to W$ into a cofibration, we use the **relative mapping cylinder**

$$Z = T_X(w) = X \cup_{X \times \Delta^1} (Y \times \Delta^1) \cup_Y W \,,$$

obtained from $T(w) = Y \times \Delta^1 \cup_Y W$ by compressing $X \times \Delta^1$ along the projection map to $X$. The front inclusion $z\colon Y \to Z$ is a finite cofibration. The back inclusion $W \to Z$ can be obtained by filling finitely many horns, hence so can its composite with $wy\colon X \to W$. This composite equals $zy$. $\square$

The non-singular simplicial case (d) of Lemma 3.2.9 is not really needed for the stable parametrized $h$-cobordism theorem, as we explain in Remark 3.5.5, after the proof of Theorem 1.2.6. However, it allows us to make some more uniform statements. We therefore include the following discussion, for completeness.

The proof of this case uses results on stellar subdivisions and their traces. Our understanding of these matters was greatly helped by a letter from Siebenmann, which explained the claims on pages 480 and 484 of his paper [Si70]. The following definition, lemmas and proof are adapted from the argument given in Siebenmann's letter.

**Definition 3.2.11.** Let $Y$ be a finite ordered simplicial complex, viewed as a simplicial set, and let $p \in |Y|$ be any point. The **star** $St(p, Y) \subset Y$ is the simplicial subset generated by the simplices $y$ such that $p$ is in the image of $|\bar{y}|\colon |\Delta^n| \to |Y|$, and the **link** $Lk(p, Y) \subset St(p, Y)$ is the simplicial subset consisting of the simplices $y$ in $St(p, Y)$ such that $p$ is not in the image of $|\bar{y}|$. Then $|St(p, Y)|$ is PL homeomorphic to the cone on $Lk(p, Y)$ with vertex $p$.

We can write $Y = St(p, Y) \cup_{Lk(p,Y)} Z$ with $Z \subset Y$. Form $\mathrm{cone}(St(p, Y))$ as a cone on $St(p, Y)$ with last vertex $v$, and define the result of **starring** $Y$ **at** $p$ as the pushout

$$S_p(Y) = \mathrm{cone}(Lk(p, Y)) \cup_{Lk(p,Y)} Z \,.$$

There is a PL homeomorphism $|\mathrm{cone}(Lk(p, Y))| \cong |St(p, Y)|$, which takes the new cone point $v$ to $p$. The induced PL homeomorphism $|S_p(Y)| \cong |Y|$ exhibits $S_p(Y)$ as a subdivision of $Y$. Let

$$T_p(Y) = \mathrm{cone}(St(p, Y)) \cup_{St(p,Y)} Z$$

be the **trace** of this (elementary stellar) subdivision. More generally, for a finite sequence of points $p_1, \ldots, p_r$ in $|Y|$, the result

$$S_{p_1 \ldots p_r}(Y) = S_{p_r}(\ldots S_{p_1}(Y) \ldots)$$

of starring $Y$ at $p_1, \ldots, p_r$, in order, is called a **stellar subdivision** of $Y$. (To make sense of this, we must identify $p_2 \in |Y|$ with its preimage under the PL homeomorphism $|S_{p_1}(Y)| \cong |Y|$, and so on.) The **trace** of this stellar subdivision is the pushout $T_{p_1 \ldots p_r}(Y)$ of the diagram

$$Y \subset T_{p_1}(Y) \supset S_{p_1}(Y) \subset \cdots \supset S_{p_1 \ldots p_{r-1}}(Y) \subset T_{p_r}(S_{p_1 \ldots p_{r-1}}(Y)) \supset S_{p_1 \ldots p_r}(Y) \,,$$

i.e., the union of the traces of the individual starring operations, glued together along the intermediate stellar subdivisions.

**Lemma 3.2.12.** *Let $p \in |Y|$. The inclusions $Y \subset T_p(Y)$ and $S_p(Y) \subset T_p(Y)$ are both simplicial expansions.*

*Let $X \subset Y$ and $p \in |X|$. The relative inclusion $T_p(X) \cup_{S_p(X)} S_p(Y) \subset T_p(Y)$ is also a simplicial expansion.*

*Proof.* It suffices to prove that the usual inclusions $St(p, Y) \subset \mathrm{cone}(St(p, Y))$, $\mathrm{cone}(Lk(p, Y)) \subset \mathrm{cone}(St(p, Y))$ and

$$\mathrm{cone}(St(p, X)) \cup_{\mathrm{cone}(Lk(p, X))} \mathrm{cone}(Lk(p, Y)) \subset \mathrm{cone}(St(p, Y))$$

are simplicial expansions. The second and third cases are special cases of the easy result that $\mathrm{cone}(X') \subset \mathrm{cone}(Y')$ is a simplicial expansion whenever $X' \subset Y'$, see the proof of Lemma 2.2.16.

To prove the first case, we temporarily work with unordered simplicial complexes. The orderings can be put back in at the end. Write $St(p, Y)$ as the join $y * L$ of the simplex $y$ containing $p$ in its interior and its link $L = Lk(y, Y)$ in $Y$. (See [RS72, 2.8(9) and 2.22(3)].) Then we can write the first inclusion as $y * L \subset v * y * L$. Here $y \subset v * y$ is an elementary simplicial expansion, from which it follows that $y * L \subset v * y * L$ is a simplicial expansion, by a straight-forward induction on the simplices in $L$. $\square$

**Lemma 3.2.13.** *Let $X \subset Y$ and $W$ be finite non-singular simplicial sets, and let $g \colon Y \to W$ be a simplicial map. The inclusion $M(Sd(g|X)) \subset M(Sd(g))$ is a simplicial expansion.*

*Proof.* By induction, we may assume that the lemma holds for all $Y$ of dimension less than $n$, and that $Y = \Delta^n \cup_{\partial \Delta^n} X$, where $\Delta^n \to Y$ is the representing map of a simplex $y \in Y_n$. Let $g(y) = \rho^*(w)$ for a degeneracy operator $\rho \colon [n] \to [m]$ and a non-degenerate simplex $w \in W_m$. Then $M(Sd(g))$ is obtained from $M(Sd(g|X))$ by attaching a copy of $M(Sd(\rho))$ along $M(Sd(\rho|\partial \Delta^n))$.

Here $M(Sd(\rho)) = N((\Delta^n)^\# \sqcup_{\rho_\#} (\Delta^m)^\#)$ is the cone on $M(Sd(\rho|\partial \Delta^n)) = N((\partial \Delta^n)^\# \sqcup_{\rho_\#} (\Delta^m)^\#)$ with vertex $(\iota_n)$, so it suffices to show that the inclusion $(\iota_m) \subset M(Sd(\rho|\partial \Delta^n))$ is a simplicial expansion. This factors as the composite of $(\iota_m) \subset Sd(\Delta^m)$, which is obviously a simplicial expansion, and the inclusion

$$Sd(\Delta^m) = M(Sd(\emptyset \to \Delta^m)) \subset M(Sd(\partial \Delta^n \to \Delta^m)),$$

which is a simplicial expansion by our inductive hypothesis, because $\partial \Delta^n$ has dimension less than $n$. Taking cones with vertex $(\iota_n)$, along any finite sequence of elementary simplicial expansions from $(\iota_m)$ to $M(Sd(\rho|\partial \Delta^n))$, then yields a finite sequence of elementary expansions from $M(Sd(\rho|\partial \Delta^n))$ union the edge $(\iota_m < \iota_n)$ to $M(Sd(\rho))$. $\square$

*Proof of Lemma 3.2.9(d).* We assume first that $y \colon X \subset Y$ is a weak homotopy equivalence of finite ordered simplicial complexes. There exists a homotopy inverse $|Y| \to |X|$, which we may assume is the identity on $|X|$. By the relative simplicial approximation theorem [Ze64, p. 40], there exists a stellar subdivision (see Definition 3.2.11 above) $Y' = S_{p_1 \dots p_r}(Y)$ of $Y$, obtained by starring $Y$ at points of $|Y|$ not in $|X|$, and a simplicial map $g \colon Y' \to X$, such that $|g|$ is homotopic to the chosen homotopy inverse relative to $|X|$. In particular, $X \subset Y'$ is a subcomplex and $g|X = id_X$.

The normal subdivision $Sd(Y')$ is a further stellar subdivision, obtained by starring $Y'$ at the barycenters of its non-degenerate simplices, in some order of decreasing dimension. Let $T_Y$ be the trace of the combined stellar subdivision $Sd(Y')$ of $Y$. It contains the trace $T_X$ of the stellar subdivision $Sd(X)$ of $X$ as a simplicial subset. Hence we have the following commutative diagram

$$
\begin{array}{ccccccccc}
X & \longrightarrow & T_X & \longleftarrow & Sd(X) & \longrightarrow & M(Sd(id_X)) & \longleftarrow & Sd(X) \\
\downarrow & & \downarrow & & \downarrow & & \downarrow & & \Big\| \\
Y & \longrightarrow & T_Y & \longleftarrow & Sd(Y') & \longrightarrow & M(Sd(g)) & \longleftarrow & Sd(X)
\end{array}
$$

where all arrows are inclusions, and $M(-)$ is the reduced mapping cylinder from Section 2.4. Note that $M(Sd(id_X)) = Sd(X) \times \Delta^1$. Let

$$
Z = T_Y \cup_{Sd(Y')} M(Sd(g))
$$

be the pushout of the lower row, and include $z \colon Y \to Z$ through $T_Y$.

We claim that the composite map $zy \colon X \to Z$ is a simplicial expansion. It suffices to check that each of the maps $X \subset T_X$, $Sd(X) \subset M(Sd(id_X))$, $M(Sd(id_X)) \subset M(Sd(g))$ and $T_X \cup_{Sd(X)} Sd(Y') \subset T_Y$ are simplicial expansions. The first and fourth of these follow by repeated application of Lemma 3.2.12. The second and third follow from Lemma 3.2.13.

We now turn to the general case. If $y \colon X \to Y$ is a finite cofibration and a weak homotopy equivalence of non-singular simplicial sets, then there exists a homotopy inverse $|Y| \to |X|$ that is the identity on $|X|$. We can write $Y \cong X \cup_{X_\alpha} Y_\alpha$ for some finite non-singular simplicial set $X_\alpha \subset X$, so that there is a finite cofibration $X_\alpha \subset Y_\alpha$ that induces $y$ upon pushout, and a retraction $|Y_\alpha| \to |X_\alpha|$ that likewise induces the homotopy inverse. Subdividing once, $Sd(X_\alpha) \subset Sd(Y_\alpha)$ is an inclusion of finite ordered simplicial complexes, and $|Sd(Y_\alpha)| \to |Sd(X_\alpha)|$ is a retraction. Proceeding as before, we get a finite cofibration $Sd(Y_\alpha) \to Z'_\alpha$ with $Sd(X_\alpha) \to Z'_\alpha$ a simplicial expansion.

Let $T_{Y_\alpha}$ be the trace of the normal subdivision of $Y_\alpha$, which contains the trace $T_{X_\alpha}$ of the normal subdivision of $X_\alpha$. Let $Z_\alpha = T_{Y_\alpha} \cup_{Sd(Y_\alpha)} Z'_\alpha$, and note that there are simplicial expansions

$$
X_\alpha \subset T_{X_\alpha} \subset T_{X_\alpha} \cup_{Sd(X_\alpha)} Z'_\alpha \subset T_{Y_\alpha} \cup_{Sd(Y_\alpha)} Z'_\alpha = Z_\alpha \,.
$$

Define $Z = X \cup_{X_\alpha} Z_\alpha$. Then we have obtained a finite cofibration $z \colon Y \to Z$ such that $zy \colon X \to Z$ is a simplicial expansion, as desired. $\quad\square$

Finally, we have the polyhedral case (e) of Lemma 3.2.9. This case is needed for the non-manifold part Theorem 1.1.8 of the stable parametrized

$h$-cobordism theorem, by way of its consequence Proposition 3.3.1(e). In principle, this case follows from the non-singular one by the possibility of triangulating polyhedra and PL maps, but enough simplification is possible in the polyhedral case that we prefer to give a direct argument. The proof uses Cohen's (non-canonical) PL mapping cylinder $C_f$ for a PL map $f \colon L \to K$, which depends on a choice of triangulation of $f$, see Remark 4.3.2.

*Proof of Lemma 3.2.9(e).* We view the given PL embedding and homotopy equivalence $\ell \colon K \to L$ as the inclusion of a subpolyhedron. There exists a homotopy inverse $L \to K$, which we can choose to be the identity on $K$. By the relative simplicial approximation theorem [Ze64], there exists a triangulation $Y$ of $L$, containing a triangulation $X$ of $K$ as a simplicial subset, and a simplicial map $g \colon Y \to X$, such that $|g|$ is homotopic to the chosen homotopy inverse relative to $K$. In particular, $|g|$ restricts to the identity on $K$. Let $M = C_{|g|} = |M(Sd(g))|$ be the PL mapping cylinder, and let $m \colon L \to M$ be the front inclusion.

Then $K \times |\Delta^1| \subset C_{|g|}$ is included as the PL mapping cylinder of $id_K$, and this inclusion is given by a finite sequence of elementary expansions. In more detail, we can inductively build $C_{|g|}$ from $K \times |\Delta^1|$ by attaching a polyhedron for each non-degenerate simplex $y$ of $Y$ not in $X$, in order of increasing simplex-dimension $n$. Let $g(y) = \rho^*(x)$, for $\rho \colon [n] \to [m]$ a degeneracy operator and $x$ a non-singular $m$-simplex in $X$. The polyhedron to be attached for $y$ is the PL mapping cylinder $C_{|\rho|}$ for $|\rho| \colon |\Delta^n| \to |\Delta^m|$, and it is attached along the PL mapping cylinder $C_{\partial|\rho|}$ of the restriction $\partial|\rho| \colon \partial|\Delta^n| \to |\Delta^m|$ of $|\rho|$ to the boundary. The map $|\rho| \colon |\Delta^n| \to |\Delta^m|$ is **dual-collapsible** in the sense of [Co67, p. 219]. Observing that $C_{|\rho|}$ is the cone on $C_{\partial|\rho|}$ with vertex the barycenter of the source simplex $|\Delta^n|$, it follows from [Co67, Thm. 7.1] that $C_{|\rho|}$ is a PL $(n+1)$-ball and that $C_{\partial|\rho|}$ is a PL $n$-ball in its boundary. Hence the inductive step in building $C_{|g|}$ from $K \times |\Delta^1|$ is precisely an elementary expansion.

More obviously, the end inclusion $K \subset K \times |\Delta^1|$ is also the composite of finitely many elementary expansions, say one for each non-degenerate simplex of any triangulation of $K$, and $m\ell$ therefore factors as the composite expansion

$$K \subset K \times |\Delta^1| \subset C_{|g|} = M \,.$$

□

To conclude the section, we will deduce Proposition 3.2.5 from the following lemma. The argument in its proof was referred to in [Wa85, Lem. 3.1.4].

**Lemma 3.2.14.** *Let $\Phi$ be a functor from simplicial sets to categories that (a) commutes with filtered colimits, and (b) takes each map obtained by filling a horn to a homotopy equivalence. Then $\Phi$ takes every weak homotopy equivalence to a homotopy equivalence.*

*Proof.* We use the fibrant replacement functor $Fx^\infty$ from Definition 3.2.10. Let

$X \to X'$ be any weak homotopy equivalence. In the commutative square

$$
\begin{array}{ccc}
X & \longrightarrow & X' \\
\downarrow & & \downarrow \\
Fx^\infty(X) & \xrightarrow{\ f\ } & Fx^\infty(X')
\end{array}
$$

the vertical weak homotopy equivalence $f_X^\infty \colon X \to Fx^\infty(X)$ is a filtered colimit of maps obtained by filling finitely many horns. Thus the two hypotheses on $\Phi$ imply that the functor $\Phi(f_X^\infty) \colon \Phi(X) \to \Phi(Fx^\infty(X))$ is a homotopy equivalence, and similarly for $X'$ in place of $X$. Considering the commutative square obtained by applying $\Phi$ to the square above, we thus are reduced to showing that the functor $\Phi(f)$ is a homotopy equivalence.

We know that $f \colon Fx^\infty(X) \to Fx^\infty(X')$ is a weak homotopy equivalence of fibrant simplicial sets, by the 2-out-of-3 property. Hence it admits a simplicial homotopy inverse $g \colon Fx^\infty(X') \to Fx^\infty(X)$ [FP90, Corollary 4.5.31(iii)]. Writing $Z = Fx^\infty(X)$ and $Z' = Fx^\infty(X')$, this means that there is a simplicial homotopy $H \colon Z \times \Delta^1 \to Z$ from $gf$ to $id_Z$, and a similar simplicial homotopy from $fg$ to $id_{Z'}$. Applying $\Phi$ we have a commutative diagram:

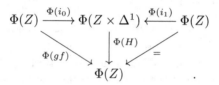

$$
\begin{array}{ccc}
\Phi(Z) & \xrightarrow{\ \Phi(i_0)\ } \Phi(Z \times \Delta^1) \xleftarrow{\ \Phi(i_1)\ } & \Phi(Z) \\
 & \Phi(gf) \searrow \quad \downarrow \Phi(H) \quad \swarrow = & \\
 & \Phi(Z) & 
\end{array}
$$

Each inclusion $i_0, i_1 \colon Z \to Z \times \Delta^1$ can be obtained by filling horns, by filling the cylinders over the non-degenerate simplices of $Z$ in some order of increasing dimension. So the hypotheses on $\Phi$ imply that the functors $\Phi(i_0)$ and $\Phi(i_1)$ are homotopy equivalences. A diagram chase then shows that $\Phi(gf)$ is a homotopy equivalence. Similarly, $\Phi(fg)$ is a homotopy equivalence, so $\Phi(f)$ is a homotopy equivalence. $\square$

*Proof of Proposition 3.2.5.* The functor $\Phi(X) = s\mathcal{C}^h(X)$ is so defined as to commute with filtered colimits, so this now follows from Proposition 3.2.2(c) and Lemma 3.2.14. $\square$

## 3.3. SOME HOMOTOPY FIBER SEQUENCES

**Proposition 3.3.1.** *(c) If $X$ is a finite simplicial set, then there is a homotopy cartesian square*

$$
\begin{array}{ccc}
s\mathcal{C}^h(X) & \longrightarrow & h\mathcal{C}^h(X) \\
\downarrow & & \downarrow \\
s\mathcal{C} & \xrightarrow{\ j\ } & h\mathcal{C}
\end{array}
$$

*with $h\mathcal{C}^h(X)$ contractible. Hence $s\mathcal{C}^h(X)$ is homotopy equivalent to the homotopy fiber of $j \colon s\mathcal{C} \to h\mathcal{C}$ at $X$. Similarly for $s\widetilde{\mathcal{C}}^h_{\bullet}(X)$.*

*(d) If $X$ is a finite non-singular simplicial set, then there is a similar homotopy cartesian square with $h\mathcal{D}^h(X)$ contractible, and $s\mathcal{D}^h(X)$ is homotopy equivalent to the homotopy fiber of $j \colon s\mathcal{D} \to h\mathcal{D}$ at $X$. Similarly for $s\widetilde{\mathcal{D}}^h_{\bullet}(X)$.*

*(e) If $K$ is a compact polyhedron, then there is a homotopy cartesian square*

$$
\begin{array}{ccc}
s\widetilde{\mathcal{E}}^h_{\bullet}(K) & \longrightarrow & h\widetilde{\mathcal{E}}^h_{\bullet}(K) \\
\downarrow & & \downarrow \\
s\widetilde{\mathcal{E}}_{\bullet} & \xrightarrow{\ \ j\ \ } & h\widetilde{\mathcal{E}}_{\bullet}
\end{array}
$$

*with $h\widetilde{\mathcal{E}}^h_{\bullet}(K)$ contractible. Hence $s\widetilde{\mathcal{E}}^h_{\bullet}(K)$ is homotopy equivalent to the homotopy fiber of $j \colon s\widetilde{\mathcal{E}}_{\bullet} \to h\widetilde{\mathcal{E}}_{\bullet}$ at $K$. Similarly for $s\mathcal{E}^h(K)$.*

*Remark 3.3.2.* The category $h\mathcal{C}^h(X)$ is contractible, because it has the initial object $id \colon X \to X$. The space of contractions of a contractible space is itself contractible. Hence we can make a contractible choice of a null-homotopy of the composite map $s\mathcal{C}^h(X) \to s\mathcal{C} \to h\mathcal{C}$, by choosing a contraction of $h\mathcal{C}^h(X)$ and composing it with the maps $s\mathcal{C}^h(X) \to h\mathcal{C}^h(X)$ and $h\mathcal{C}^h(X) \to h\mathcal{C}$. Similar remarks apply in the other cases.

In order to make sure that (0.4) commutes, as a diagram of homotopy fiber sequences, it is convenient to note that these null-homotopies can all be chosen compatibly, because the preferred contraction of $h\mathcal{C}^h(X)$ obtained from the initial object is strictly compatible with the corresponding preferred contractions of $h\mathcal{D}^h(X)$ and $h\widetilde{\mathcal{E}}^h_{\bullet}(|X|)$.

The proof of Proposition 3.3.1 will be given after four preparatory lemmas. The reader is encouraged to first peruse Section 3.6 for a review of the categorical definitions concerning right fibers, transition maps and so on, especially as they apply for simplicial functors between simplicial categories.

**Lemma 3.3.3.** *(c) If $X$ is a finite simplicial set, then the full embedding*

$$
s\mathcal{C}^h(X) \xrightarrow{\ \simeq\ } X/j
$$

*is a homotopy equivalence. Similarly for $s\widetilde{\mathcal{C}}^h_{\bullet}(X)$.*

*(d) If $X$ is a finite non-singular simplicial set, then the full embedding*

$$
s\mathcal{D}^h(X) \xrightarrow{\ \simeq\ } X/j
$$

*is a homotopy equivalence. Similarly for $s\widetilde{\mathcal{D}}^h_{\bullet}(X)$.*

*(e) If $K$ is a compact polyhedron, then the full embedding*

$$
s\widetilde{\mathcal{E}}^h_{\bullet}(K) \xrightarrow{\ \simeq\ } ([0], K)/j
$$

*is a homotopy equivalence. Similarly for $s\mathcal{E}^h(K)$.*

*Proof.* The embedding identifies $s\mathcal{C}^h(X)$ with the full subcategory of $X/j$ generated by the objects $y \colon X \to Y$ that are not just weak homotopy equivalences of finite simplicial sets, but also cofibrations. We define a homotopy inverse functor $X/j \to s\mathcal{C}^h(X)$ that takes $y \colon X \to Y$ to the finite cofibration

$$X \xrightarrow{(i,y)} \mathrm{cone}(X) \times Y \,.$$

The projection $pr \colon \mathrm{cone}(X) \times Y \to Y$ is a simple map, under $X$, and simultaneously defines the two required natural transformations to the identity endofunctor of $s\mathcal{C}^h(X)$ and $X/j$, respectively. The proofs in the non-singular and polyhedral cases are identical.

In the case of simplicial categories, the embedding identifies $s\widetilde{\mathcal{E}}^h_\bullet(K)$ with the full subcategory of $([0], K)/j$ generated in simplicial degree $q$ by the objects

$$pr \colon K \times |\Delta^q| \xrightarrow{e} E \xrightarrow{\pi} |\Delta^q|$$

where not only is $\pi$ a PL Serre fibration and $e$ a PL homotopy equivalence of compact polyhedra, but $e$ is also an embedding. Again we can define a homotopy inverse functor $([0], K)/j \to s\widetilde{\mathcal{E}}^h_\bullet(K)$ by taking an object as above to

$$pr \colon K \times |\Delta^q| \xrightarrow{(i,e)} \mathrm{cone}(K \times |\Delta^q|) \times E \xrightarrow{\pi \circ pr} |\Delta^q| \,.$$

It is clear that $(i, e)$ is a PL embedding and a homotopy equivalence, and that $\pi \circ pr$ is the composite of two Serre fibrations, and therefore again a Serre fibration. The projection map

$$\mathrm{cone}(K \times |\Delta^q|) \times E \xrightarrow{pr} E$$

is a simple map of Serre fibrations, and commutes with the embeddings of $K \times |\Delta^q|$ and the projections to $|\Delta^q|$, so it defines the two natural transformations to the identity needed to exhibit the two given functors as mutual homotopy inverses. The proof in the case of general, or non-singular, simplicial sets is identical. $\square$

*Remark 3.3.4.* It is not clear that the analog of Lemma 3.3.3 holds for the relative PL bundle categories $s\mathcal{D}^h_\bullet(X)$ and $s\mathcal{E}^h_\bullet(K)$, where the PL Serre fibrations are required to be PL bundles relative to a given product subbundle. In the notation of the previous proof, $\pi \circ pr \colon \mathrm{cone}(K \times |\Delta^q|) \times E \to |\Delta^q|$ will be PL locally trivial when $\pi \colon E \to |\Delta^q|$ is a PL bundle, but there is no guarantee that its PL local trivializations can be arranged to be locally constant on the product subbundle given by the embedding $(i, e) \colon K \times |\Delta^q| \to \mathrm{cone}(K \times |\Delta^q|) \times E$, as demanded in Definition 3.1.7.

**Lemma 3.3.5.** *If $j\colon s\mathcal{C} \to h\mathcal{C}$ is the inclusion functor, and $f\colon X \to X'$ is a morphism in $h\mathcal{C}$, then the transition map*

$$f^*\colon X'/j \xrightarrow{\simeq} X/j$$

*is a homotopy equivalence. Similarly for $j\colon s\mathcal{D} \to h\mathcal{D}$ and for $j\colon s\mathcal{E} \to h\mathcal{E}$.*

*If $j\colon s\widetilde{\mathcal{E}}_\bullet \to h\widetilde{\mathcal{E}}_\bullet$ is the inclusion functor, and $f\colon K \to K'$ is a morphism in $h\widetilde{\mathcal{E}}_\bullet$ in simplicial degree 0, then the transition map*

$$([0], f)^*\colon ([0], K')/j \xrightarrow{\simeq} ([0], K)/j$$

*is a homotopy equivalence. Similarly for $j\colon s\widetilde{\mathcal{C}}_\bullet \to h\widetilde{\mathcal{C}}_\bullet$ and for $j\colon s\widetilde{\mathcal{D}}_\bullet \to h\widetilde{\mathcal{D}}_\bullet.$*

*Proof.* We use the cone-product construction to factor the weak homotopy equivalence $f\colon X \to X'$ as the composite

$$X \xrightarrow{(i,f)} \mathrm{cone}(X) \times X' \xrightarrow{pr} X'$$

of the front inclusion $(i, f)$ and the projection $pr$. The back inclusion $s\colon X' \to \mathrm{cone}(X) \times X'$ is a section to $pr$, and all of these maps are weak homotopy equivalences. The induced transition maps satisfy $f^* = (i, f)^* pr^*$ and $s^* pr^* = id$, so it suffices to prove the lemma for the cofibrations $(i, f)$ and $s$, i.e., we may assume from the outset that $f$ is a cofibration (and a weak homotopy equivalence). Then we have a commutative square

$$
\begin{array}{ccc}
s\mathcal{C}^h(X') & \xrightarrow{\;f^*\;} & s\mathcal{C}^h(X) \\
\downarrow & & \downarrow \\
X'/j & \xrightarrow{\;f^*\;} & X/j
\end{array}
$$

where the vertical functors are homotopy equivalences by Lemma 3.3.3 and the upper horizontal functor $f^*$ is a homotopy equivalence by Proposition 3.2.3. Hence also the lower horizontal functor $f^*$ is a homotopy equivalence, as asserted. The proofs in the non-singular and polyhedral cases, as well as for the simplicial categories, are the same. $\square$

**Lemma 3.3.6.** *Let $\pi\colon Z \to \Delta^q$ be an object of $h\widetilde{\mathcal{D}}_\bullet$ in simplicial degree $q$, and let $t\colon P = \Phi \times \Delta^q \to Z$ be a trivialization up to a simple map. Then the transition map*

$$([q], t)^*\colon ([q], Z)/j \xrightarrow{\simeq} ([q], P)/j$$

*is a homotopy equivalence. Similarly for $h\widetilde{\mathcal{C}}_\bullet$ and for $h\widetilde{\mathcal{E}}_\bullet.$*

*Proof.* As we will recall in Definition 3.6.6, the right fiber $([q], Z)/j$ is the simplicial category

$$[n] \longmapsto \coprod_{\gamma\colon [n]\to[q]} \gamma^* Z/j_n \,,$$

where $\gamma$ ranges over the indicated morphisms in $\Delta$, and $\gamma^* Z$ is an object in the target of $j_n \colon s\widetilde{D}_n \to h\widetilde{D}_n$.

We first show that the (non-simplicial) transition map

$$t^* \colon Z/j_q \to P/j_q$$

of right fibers of $j_q \colon s\widetilde{D}_q \to h\widetilde{D}_q$ is a homotopy equivalence. We cannot define a covariant functor $t_* \colon P/j_q \to Z/j_q$ on an object

$$pr \colon P \xrightarrow{w} W \xrightarrow{\pi} \Delta^q$$

as the pushout of $Z \xleftarrow{t} P \xrightarrow{w} W$, because this construction will usually only preserve weak homotopy equivalences when the simple map $t$ is a cofibration, and this only happens when $t$ is an isomorphism. However, we can use the cone-product construction from Section 2.4 to form the pushout of the two cofibrations

$$\mathrm{cone}(P) \times Z \xleftarrow{(i,t)} P \xrightarrow{(i,w)} \mathrm{cone}(P) \times W \,,$$

with $Z$ mapping to $* \times Z$ on the left. Here $*$ denotes the vertex of $\mathrm{cone}(P)$. We call this pushout $t_!(W)$, for the duration of this proof. It is a Serre fibration over $\Delta^q$ after geometric realization, by the fiber gluing lemma for Serre fibrations.

With this notation, $t^* t_!(W)$ is the same simplicial set as $t_!(W)$, but it is viewed as an object under $P$ via $t$ and the stated inclusion of $Z$. It receives a simple map (using $t \colon P \to Z$ on the left) from the pushout of the two cofibrations

$$\mathrm{cone}(P) \times P \xleftarrow{(i,id)} P \xrightarrow{(i,w)} \mathrm{cone}(P) \times W \,,$$

which maps simply (using $pr \colon \mathrm{cone}(P) \times P \to P$) to

$$\mathrm{cone}(P) \times W \,.$$

This projects by a simple map to $W$. All these simple maps are compatible with the structure maps from $P$, so these pushouts define a chain of natural transformations of endofunctors of $P/j_q$, which relate $t^* t_!$ to the identity.

On the other hand, if

$$pr \colon Z \xrightarrow{w} W \xrightarrow{\pi} \Delta^q$$

is an object of $Z/j_q$, then $t_! t^*(W)$ is the pushout of the two cofibrations

$$\mathrm{cone}(P) \times Z \xleftarrow{(i,t)} P \xrightarrow{(i,wt)} \mathrm{cone}(P) \times W \,,$$

as an object under $Z$. There is a natural simple map (using $t \colon P \to Z$ in all places) to the pushout of the two cofibrations

$$\mathrm{cone}(Z) \times Z \xleftarrow{(i,id)} Z \xrightarrow{(i,w)} \mathrm{cone}(Z) \times W \,,$$

and a second simple map (using $pr$: $\mathrm{cone}(Z) \times Z \to Z$) to

$$\mathrm{cone}(Z) \times W \,.$$

Again, this projects by a simple map to $W$. All these simple maps are compatible with the structure maps from $Z$, so there is a natural transformation of endofunctors of $Z/j_q$, from $t_!t^*$ to the identity. This proves that $t^* \colon Z/j_q \to P/j_q$ is a homotopy equivalence.

More generally, for each morphism $\gamma \colon [n] \to [q]$ in $\Delta$ there is a pulled-back simple map $\gamma^*t \colon \gamma^*P \to \gamma^*Z$, with $\gamma^*P = \Phi \times \Delta^n$. By the previous argument, applied over $\Delta^n$ in place of $\Delta^q$, the functor $(\gamma^*t)^* \colon \gamma^*Z/j_n \to \gamma^*P/j_n$ is a homotopy equivalence. The disjoint union over these $\gamma \colon [n] \to [q]$ of the functors $(\gamma^*t)^*$ is, by definition, the degree $n$ part of $([q],t)^*$. Hence the latter is also a homotopy equivalence, by the realization lemma (which we recall in Proposition 3.6.11).

The constructions used, i.e., cones, products and pushouts along pairs of cofibrations, are all available in the general simplicial and the polyhedral contexts, so the same proof applies in these cases.  $\square$

**Lemma 3.3.7.** *Let $f \colon Z \to Z'$ be a morphism in $h\widetilde{\mathcal{D}}_{\bullet}$ in simplicial degree $q$, and let $\alpha \colon [p] \to [q]$ be a morphism in $\Delta$. To verify that all transition maps*

$$([q],f)^* \colon ([q],Z')/j \to ([q],Z)/j$$

*(of the first kind) and*

$$\alpha_* \colon ([p], \alpha^*Z)/j \to ([q],Z)/j$$

*(of the second kind) are homotopy equivalences, it suffices to do so for the transition maps $([0],f)^*$ of morphisms $f$ in simplicial degree $0$. Similarly for $h\widetilde{\mathcal{C}}_{\bullet}$ and for $h\widetilde{\mathcal{E}}_{\bullet}$.*

*Proof.* With $\Phi = Sd(\pi)^{-1}(\beta)$ the preimage of the barycenter ($\Phi$ for fiber, see Definition 2.7.1), we have a natural simple map

$$t \colon P = \Phi \times \Delta^q \to Z$$

over $\Delta^q$. See Proposition 2.7.6. Here $P = \epsilon^*\Phi$ for the unique map $\epsilon \colon [q] \to [0]$ in $\Delta$, and similarly when decorated with primes. Letting $\varphi \colon \Phi \to \Phi'$ be the restriction of $Sd(f)$ to $\Phi$, we have a commutative diagram

$$
\begin{array}{ccccc}
([q],Z')/j & \xrightarrow{\;t'^*\;} & ([q],P')/j & \xrightarrow{\;\epsilon_*\;} & ([0],\Phi')/j \\
\downarrow{\scriptstyle ([q],f)^*} & & \downarrow{\scriptstyle ([q],\epsilon^*\varphi)^*} & & \downarrow{\scriptstyle ([0],\varphi)^*} \\
([q],Z)/j & \xrightarrow{\;t^*\;} & ([q],P)/j & \xrightarrow{\;\epsilon_*\;} & ([0],\Phi)/j \,.
\end{array}
$$

The left hand square commutes by the naturality of $t$ with respect to $f$. By Lemma 3.3.6 the maps $t^*$ and $t'^*$ are homotopy equivalences. The maps $\epsilon_*$ are always homotopy equivalences, as in the proof of Addendum 3.6.10. Hence $([q], f)^*$ is a homotopy equivalence if and only if the degree 0 transition map $([0], \varphi)^*$ is a homotopy equivalence. Thus if the transition maps of the first kind in simplicial degree 0 are homotopy equivalences, then so are all the transition maps of the first kind.

Let $\eta\colon [0] \to [q]$ be any morphism from $[0]$ in $\Delta$, and let $\tau = \eta^* t\colon \Phi = \eta^* P \to \eta^* Z$. Then in the commutative diagram

$$
\begin{array}{ccc}
([0], \eta^* Z)/j & \xrightarrow{\ \eta_*\ } & ([q], Z)/j \\
\Big\downarrow{\scriptstyle ([0],\tau)^*} & & \Big\downarrow{\scriptstyle ([q],t)^*} \\
([0], \Phi)/j & \xrightarrow{\ \eta_*\ } & ([q], P)/j \xrightarrow{\ \epsilon_*\ } ([0], \Phi)/j
\end{array}
$$

the lower composite map is the identity because $\epsilon\eta = id$, the map $\epsilon_*$ is a homotopy equivalence as before, and $([q], t)^*$ is a homotopy equivalence by Lemma 3.3.6. Thus the upper horizontal map $\eta_*$ is a homotopy equivalence if and only if the degree 0 transition map $([0], \tau)^*$ is one.

In other words, if all transition maps of the first kind in simplicial degree 0 are homotopy equivalences, then so are all the transition maps of the second kind for morphisms $\eta$ from $[0]$ in $\Delta$. Now let $\alpha\colon [p] \to [q]$ and $\eta\colon [0] \to [p]$ be any morphisms in $\Delta$. Then we have just shown that $\eta_*$ and $(\alpha\eta)_*$ are homotopy equivalences, which implies that also the general transition map $\alpha_*$ of the second kind is a homotopy equivalence.

The general simplicial case follows by the same proof.

The polyhedral case also follows by the same argument, in view of the possibility of triangulating a PL map $f\colon E \to E'$ of PL Serre fibrations over $|\Delta^q|$, by a map $Y \to Y'$ of Serre fibrations over a simplicially collapsible base $X$. The use of Proposition 2.7.6 must be replaced by an appeal to Proposition 2.7.7, the fiber $\Phi$ must be replaced by the polyhedron $|L_X|$ of that proposition, and similarly for $\Phi'$. Otherwise, the argument proceeds in the same way. $\square$

*Proof of Proposition 3.3.1.* The first part of Lemma 3.3.5 verifies the hypothesis of Quillen's Theorem B for the functor $j\colon s\mathcal{C} \to h\mathcal{C}$. Hence the right hand square in the following diagram is homotopy cartesian.

$$
\begin{array}{ccc}
s\mathcal{C}^h(X) \longrightarrow & X/j \longrightarrow & s\mathcal{C} \\
\Big\downarrow & \Big\downarrow & \Big\downarrow{\scriptstyle j} \\
h\mathcal{C}^h(X) \longrightarrow & X/h\mathcal{C} \longrightarrow & h\mathcal{C}
\end{array}
$$

The full embedding $s\mathcal{C}^h(X) \to X/j$ is a homotopy equivalence, by Lemma 3.3.3. The categories $h\mathcal{C}^h(X)$ and $X/h\mathcal{C}$ each have an initial object, and are therefore

contractible. It follows that also the outer rectangle is homotopy cartesian.
The same argument applies for $j: s\mathcal{D} \to h\mathcal{D}$ at $X$ and for $j: s\mathcal{E} \to h\mathcal{E}$ at $K$.

The remaining part of Lemma 3.3.5, and Lemma 3.3.7, verify the hypotheses
of Theorem B$'$ for the simplicial functor $j: s\widetilde{\mathcal{E}}_\bullet \to h\widetilde{\mathcal{E}}_\bullet$. Hence the right hand
square in the following diagram is homotopy cartesian.

$$
\begin{array}{ccccc}
s\widetilde{\mathcal{E}}_\bullet^h(K) & \longrightarrow & ([0], K)/j & \longrightarrow & s\widetilde{\mathcal{E}}_\bullet \\
\downarrow & & \downarrow & & \downarrow{\scriptstyle j} \\
h\widetilde{\mathcal{E}}_\bullet^h(K) & \longrightarrow & ([0], K)/h\widetilde{\mathcal{E}}_\bullet & \longrightarrow & h\widetilde{\mathcal{E}}_\bullet
\end{array}
$$

The full embedding $s\widetilde{\mathcal{E}}_\bullet^h(K) \to ([0], K)/j$ is a homotopy equivalence, by
the remaining part of Lemma 3.3.3. The simplicial categories $h\widetilde{\mathcal{E}}_\bullet^h(K)$ and
$([0], K)/h\widetilde{\mathcal{E}}_\bullet$ each have an initial object in each degree, and are therefore contractible. It follows that also the outer rectangle is homotopy cartesian. The
same argument applies for $j: s\widetilde{\mathcal{C}}_\bullet \to h\widetilde{\mathcal{C}}_\bullet$ and for $j: s\widetilde{\mathcal{D}}_\bullet \to h\widetilde{\mathcal{D}}_\bullet$ at $X$.    $\square$

## 3.4. POLYHEDRAL REALIZATION

**Definition 3.4.1.** By a **triangulation** of a compact topological space $K$ we
will mean a homeomorphism $h: |X| \xrightarrow{\cong} K$, where $X$ is a finite non-singular
simplicial set. A second triangulation $h': |X'| \xrightarrow{\cong} K$ is a **linear subdivision** of
the first if each simplex of $X'$ maps (affine) linearly into a simplex of $X$, in terms
of barycentric coordinates, after geometric realization. In other words, for each
simplex $x' \in X'_m$ there is a simplex $x \in X_n$ and a linear map $\ell: |\Delta^m| \to |\Delta^n|$
such that the following diagram commutes.

$$
\begin{array}{ccc}
|\Delta^m| & \xrightarrow{\quad\ell\quad} & |\Delta^n| \\
{\scriptstyle |\bar{x}'|}\downarrow & & \downarrow{\scriptstyle |\bar{x}|} \\
|X'| \xrightarrow[\cong]{h'} & K \xleftarrow[\cong]{h} & |X|
\end{array}
$$

(If desired, one may restrict attention to non-degenerate simplices $x$ and $x'$
in this definition.) As regards the ordering of vertices, we shall furthermore
assume that if for some $0 \le i, j \le m$ the linear map $\ell: |\Delta^m| \to |\Delta^n|$ takes the
$i$-th vertex of $|\Delta^m|$ to the boundary of $|\Delta^n|$ and the $j$-th vertex to the interior
of $|\Delta^n|$, then $i < j$. In other words, the interior vertices are to have the bigger
numbering.

Two triangulations are **equivalent** if they admit a common linear subdivision, and a **compact polyhedron** is a compact space $K$ together with
a **PL structure**, i.e., an equivalence class of triangulations. A **PL map**
$K \to L$ of compact polyhedra is one of the form $h'|f|h^{-1}$, for some trian-
gulations $h: |X| \xrightarrow{\cong} K$ and $h': |Y| \xrightarrow{\cong} L$ in the respective PL structures, and

$f\colon X \to Y$ a simplicial map. We let $r\colon \mathcal{D} \to \mathcal{E}$ be the **polyhedral realization** functor, which takes a finite non-singular simplicial set $X$ to its geometric realization $|X|$, with the PL structure generated by the identity triangulation $id\colon |X| \overset{\cong}{\to} |X|$.

*Remark 3.4.2.* These are the classical definitions, except that one usually asks that $X$ is a finite simplicial complex instead of a finite non-singular simplicial set [FP90, §3.4]. However, the barycentric subdivision $X'$ of any finite simplicial complex $X$ is a finite ordered simplicial complex, and the normal subdivision $Sd(X)$ of any finite non-singular simplicial set is a finite simplicial set in which each non-degenerate simplex is uniquely determined by its set of vertices (= faces in simplicial degree 0). We can canonically identify finite ordered simplicial complexes with the latter kind of simplicial sets. It follows that each PL structure in the classical sense intersects nontrivially with a unique PL structure in the present sense, and *vice versa*, so the definitions are equivalent.

Explicitly, the classical PL structure generated by a triangulation $h\colon |X| \overset{\cong}{\to} K$ for a simplicial complex $X$, corresponds to the PL structure generated by the barycentrically subdivided triangulation $h'\colon |X'| \overset{\cong}{\to} K$, where $X'$ is viewed as a non-singular simplicial set. Conversely, the current PL structure generated by a triangulation $h\colon |X| \overset{\cong}{\to} K$ for a non-singular simplicial set $X$, corresponds to the classical PL structure generated by the normally subdivided triangulation $h \circ h_X\colon |Sd(X)| \overset{\cong}{\to} K$, where $Sd(X)$ is viewed as a simplicial complex.

*Remark 3.4.3.* For a given triangulation $h\colon |X| \overset{\cong}{\to} K$, the normal subdivision $Sd(X)$ provides a subdivided triangulation

$$h' = h \circ h_X\colon |Sd(X)| \overset{\cong}{\to} K ,$$

where $h_X\colon |Sd(X)| \to |X|$ is the canonical (but not natural) homeomorphism, as in Theorem 2.3.1. In particular, the subdivided triangulation satisfies our convention on the ordering of vertices. By [Hu69, Cor. 1.6], each linear subdivision $X'$ of a finite simplicial complex $X$ admits a further linear subdivision $X''$ that is isomorphic to an iterated normal subdivision $Sd^r(X)$, for some $r$.

**Proposition 3.4.4.** *The polyhedral realization functors* $r\colon s\mathcal{D}_\bullet \to s\mathcal{E}_\bullet$ *and*

$$r\colon s\widetilde{\mathcal{D}}_\bullet \to s\widetilde{\mathcal{E}}_\bullet ,$$

*as well as their analogs* $r\colon h\mathcal{D}_\bullet \to h\mathcal{E}_\bullet$ *and* $r\colon h\widetilde{\mathcal{D}}_\bullet \to h\widetilde{\mathcal{E}}_\bullet$, *are homotopy equivalences.*

In contrast to Proposition 3.5.1, we do not claim that these simplicial functors are degreewise homotopy equivalences. In particular, we do not claim to know that $r\colon s\mathcal{D} \to s\mathcal{E}$ is a homotopy equivalence.

*Proof.* We will concentrate on the case of the functor $r\colon s\widetilde{\mathcal{D}}_\bullet \to s\widetilde{\mathcal{E}}_\bullet$. The other cases differ only by trivial modifications. The claim is that the bisimplicial map of nerves

$$[k], [q] \longmapsto (s_k\widetilde{\mathcal{D}}_q \overset{r}{\to} s_k\widetilde{\mathcal{E}}_q)$$

is a weak homotopy equivalence, where $s_k \widetilde{\mathcal{D}}_q$ is the simplicial degree $k$ part of the nerve of the category $s\widetilde{\mathcal{D}}_q$, and similarly for $s_k \widetilde{\mathcal{E}}_q$. By the realization lemma (see Proposition 3.6.11) it will be enough to show that the map of simplicial sets

$$r \colon s_k \widetilde{\mathcal{D}}_\bullet \to s_k \widetilde{\mathcal{E}}_\bullet$$

is a weak homotopy equivalence for each $k$. Here $s_k \widetilde{\mathcal{D}}_\bullet$ is the simplicial set with $q$-simplices $s_k \widetilde{\mathcal{D}}_q$, i.e., the sequences of $k$ composable simple maps

$$Z_0 \xrightarrow{f_1} Z_1 \xrightarrow{f_2} \ldots \xrightarrow{f_k} Z_k$$

of Serre fibrations over $\Delta^q$, where each $Z_i$ is a finite non-singular simplicial set, and similarly for $s_k \widetilde{\mathcal{E}}_\bullet$.

For any choice of base point in $|s_k \widetilde{\mathcal{D}}_\bullet|$, each element in the relative homotopy group or set $\pi_i(r)$ can be represented by the geometric realization of a pair of maps $(e, d)$ as in the following lemma, with $A \subset B$ a suitable subdivision of the pair $\partial\Delta^i \subset \Delta^i$.

**Lemma 3.4.5.** *Let $A \subset B$ be a pair of finite non-singular simplicial sets and*

$$
\begin{array}{ccc}
A & \xrightarrow{\ d\ } & s_k \widetilde{\mathcal{D}}_\bullet \\
\downarrow & & \downarrow{\scriptstyle r} \\
B & \xrightarrow{\ e\ } & s_k \widetilde{\mathcal{E}}_\bullet
\end{array}
$$

*a commutative square. Then there exists a finite non-singular simplicial set $T$ containing $A$, a weak homotopy equivalence $h \colon T \to B$ that is the identity on $A$, and a map*

$$g \colon T \to s_k \widetilde{\mathcal{D}}_\bullet$$

*such that $g|A = d$ and $eh$ is homotopic to $rg$ relative to $A$.*

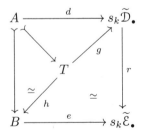

Granting this, we can now complete the proof of Proposition 3.4.4. Right composition with the homotopy equivalence $(|h|, id) \colon (|T|, |A|) \to (|B|, |A|)$ obtained from the lemma induces a bijection on homotopy classes of maps, which takes the homotopy class of $(|e|, |d|)$ to that of $(|eh|, |d|)$. Also by the lemma, the latter is homotopic to $(|rg|, |d|)$, which represents 0. Thus all the relative homotopy groups $\pi_i(r)$ are 0, and therefore $r$ is a weak homotopy equivalence.   $\square$

*Proof of Lemma 3.4.5.* We translate into more concrete terms what it means to have a map $e \colon B \to s_k \widetilde{\mathcal{E}}_{\bullet}$. Suppose first that $k = 0$. A $q$-simplex of $s_0 \widetilde{\mathcal{E}}_{\bullet}$ is just a PL Serre fibration $\pi \colon E \to |\Delta^q|$. If we are given a map $e \colon B \to s_0 \widetilde{\mathcal{E}}_{\bullet}$ we obtain a space over $|B|$, by gluing over the simplices of $B$. If $B$ is finite and non-singular then this space is a polyhedron, $P$ say, and the map $\pi \colon P \to |B|$ is piecewise-linear. To characterize $P$ it suffices to say that if $b$ is a $q$-simplex of $B$, then the pullback by $|\bar{b}| \colon |\Delta^q| \to |B|$ of $P$ is the polyhedron $e(b)$ over $|\Delta^q|$. By the base gluing lemma for Serre fibrations (Lemma 2.7.12), $\pi \colon P \to |B|$ is also a Serre fibration, and conversely, any PL Serre fibration $\pi \colon P \to |B|$ arises from a unique map $e \colon B \to s_k \widetilde{\mathcal{E}}_{\bullet}$, up to isomorphism. More generally, we obtain in this fashion a correspondence between maps $e \colon B \to s_k \widetilde{\mathcal{E}}_{\bullet}$ and sequences of $k$ simple maps

$$(3.4.6) \qquad\qquad P_0 \to P_1 \to \ldots \to P_k$$

of PL Serre fibrations over $|B|$, where each $P_i$ is a compact polyhedron.

For $A \subset B$, the restriction $e|A \colon A \to s_k \widetilde{\mathcal{E}}_{\bullet}$ then corresponds to the restricted sequence of $k$ simple maps

$$(3.4.7) \qquad\qquad Q_0 \to Q_1 \to \ldots \to Q_k$$

of PL Serre fibrations over $|A|$, where each $Q_i = |A| \times_{|B|} P_i$ is the restriction of $P_i \to |B|$ to $|A| \subset |B|$. A map $d \colon A \to s_k \widetilde{\mathcal{D}}_{\bullet}$ with $rd = e|A$ then corresponds to a triangulation of the sequence above, i.e., a sequence of $k$ simple maps

$$X_0 \to X_1 \to \ldots \to X_k$$

of Serre fibrations over $A$, where each $X_i$ is a finite non-singular simplicial set, whose polyhedral realization is the sequence (3.4.7). Here we again use the base gluing lemma for Serre fibrations, to equate the Serre fibrations over $A$ with the simplicial maps that restrict to a Serre fibration over each simplex of $A$.

Now we appeal to the following (sub-)lemma.

**Lemma 3.4.8.** *The sequence of polyhedra and PL maps*

$$P_0 \to P_1 \to \ldots \to P_k \to |B|$$

*may be triangulated by a diagram*

$$Y_0 \to Y_1 \to \ldots \to Y_k \xrightarrow{\pi} B'$$

*of finite non-singular simplicial sets, in such a way that:*
  (1) *the triangulation $B'$ of $|B|$ is a linear subdivision of $B$; and*
  (2) *for every $i$, $Q_i \subset P_i$ is a subcomplex with an induced triangulation $X_i' \subset Y_i$ that is a linear subdivision of its original triangulation $X_i$.*

*Proof.* By triangulation theory, e.g. [Hu69, Thm. 1.11] or [RS72, Thm. 2.15], it is always possible to triangulate a sequence of PL maps of polyhedra, and moreover the triangulations can be chosen to be fine, in the sense that they refine given triangulations and contain given subpolyhedra as subcomplexes. This is because the polyhedra, subpolyhedra and simplices of the given triangulations can be assembled in a **one-way tree**. Our convention on orderings may be forced true, if necessary, by an additional (pseudo-)barycentric subdivision. □

Returning to the proof of Lemma 3.4.5, we let

$$|T| = |A \times \Delta^1| \cup_{|A|} |B|$$

be the mapping cylinder of the embedding $|A| \subset |B|$, and define

$$R_i = Q_i \times |\Delta^1| \cup_{Q_i} P_i$$

for each $i$. We obtain a sequence of $k$ simple maps

(3.4.9)                                $R_0 \to R_1 \to \dots \to R_k$

of PL Serre fibrations over $|T|$, namely that obtained from (3.4.6) by pullback along the cylinder projection $p \colon |T| \to |B|$.

The polyhedron $|T|$ will be triangulated as $T = A'' \cup_{A'} B'$, where $B'$ is the linear subdivision of $B$ from Lemma 3.4.8, $A'$ is the induced subdivision of $A$, and $A''$ is a triangulation of the cylinder $|A \times \Delta^1|$ that extends the triangulations $A$ and $A'$ at the two ends. To be specific, $A''$ is obtained by inductively **starring** each of the convex **blocks** $|\Delta^m \times \Delta^1|$ from some interior point, with one block for each non-degenerate simplex $\bar{a} \colon \Delta^m \to A$, in order of increasing dimension. In other words, we have inductively triangulated the boundary of this block, and triangulate the whole block as the cone on its boundary, with any interior point as the cone vertex. As regards ordering, we always make the cone vertex the last vertex of the new simplices.

The sequence (3.4.9) now represents the map that will be $rg \colon T \to s_k \widetilde{\mathcal{E}}_\bullet$. To define the lifting $g \colon T \to s_k \widetilde{\mathcal{D}}_\bullet$ we must triangulate the $R_i$ and the maps between them in a certain way. Lemma 3.4.8 provides triangulations $Y_i$ over $B'$ of the parts $P_i \subset R_i$. We triangulate the subpolyhedra $Q_i \subset R_i$ (at the front ends) by the original triangulations $X_i$ over $A$. This ensures that $g|A = d$. Finally, the blocks $|\Delta^n \times \Delta^1|$ of $Q_i \times |\Delta^1|$ are triangulated inductively, in order of increasing $n$, by starring at suitable interior points. Here some care must be taken to ensure that these interior points are chosen compatibly with the maps in (3.4.7) and the bundle projection to $|A|$. This is done by triangulating from $Q_k$ back towards $Q_0$, and noting that for any surjective linear map $|\Delta^n| \to |\Delta^m|$ the interior of $|\Delta^n|$ maps onto the interior of $|\Delta^m|$. As before, the new cone vertex is always numbered as the last vertex.

To complete the proof of the lemma, we must define the weak homotopy equivalence $h \colon T \to B$, and show that $rg$ is homotopic to $eh$, in both cases relative to $A$. The map $h$ will be a variant last vertex map $d'$ from $T = A'' \cup_A B'$ to the block complex $A \times \Delta^1 \cup_A B$ that it subdivides (with $A \times \Delta^1$ decomposed into blocks $\Delta^m \times \Delta^1$, rather than into simplices), followed by the cylinder projection $p$ to $B$. Instead of developing a theory of last vertex maps for block complexes, extending Section 2.2, we just describe the composite map $h = pd'$ directly.

Each 0-simplex $v \in T_0$ is a point of $|T|$, which maps under $p \colon |T| \to |B|$ to the interior of a unique non-degenerate simplex $b(v)$ of $B$, often called the **support** of $p(v)$. We define $h(v) \in B_0$ to be the last vertex of this simplex $b(v)$, and note that after geometric realization, the convex linear path from $p(v)$ to $h(v)$ lies within $b(v)$. We then extend $h$ linearly on each simplex, as we now explain. Let $t \in T_n$ be any $n$-simplex, with vertices $v_0, \dots, v_n$, in order. First, suppose that $t$ lies in $B' \subset T$. Since $B'$ is a linear subdivision of $B$, $t$ maps linearly to some non-degenerate simplex $b$ in $B$, after geometric realization. Second, suppose that $t$ lies in $A'' \subset T$. Since $A''$ is a linear subdivision of the block complex $A \times \Delta^1$, whose blocks project linearly under $p$ to the simplices of $B$, again $t$ maps linearly under $p$ to some non-degenerate simplex $b$ in $B$, all after geometric realization. If $b$ has dimension $m$, then by our convention on orderings, there is a well-defined morphism $\alpha \colon [n] \to [m]$ in $\Delta$ such that $h(v_i)$ is the $\alpha(i)$-th vertex of $b$, for each $i$, and then we let $h(t) = \alpha^*(b) \in B_n$. This defines the simplicial map

$$h \colon T = A'' \cup_{A'} B' \to B.$$

(In the special case when $B' = Sd(B)$ is the normal subdivision, the restriction to $B' \subset T$ of the map $h$ just defined equals the usual last vertex map $d_B$.) On $A \subset A''$ the cylinder projection $p$ realizes the inclusion $|A| \subset |B|$, so $p(v) = h(v)$ for the vertices $v \in A_0$, and the restriction to $A \subset T$ of $h$ equals the inclusion $A \subset B$.

The convex linear paths from $p(v)$ to $h(v)$ define a homotopy $p \simeq |h| \colon |T| \to |B|$ that is linear on each simplex of $T$, and is constant on $|A|$. Thus $h$ is a weak homotopy equivalence. Finally, $rg$ represents the sequence (3.4.9) obtained by pullback from (3.4.6) along $p$, and $eh$ represents the sequence obtained by pullback along $|h|$. Pullback of (3.4.6) along the homotopy $|T \times \Delta^1| \to |B|$ from $p$ to $|h|$, which is linear on each simplex, then defines a sequence of $k$ simple maps of PL Serre fibrations over $|T \times \Delta^1|$. The corresponding map $T \times \Delta^1 \to s_k \widetilde{\mathcal{E}}_{\bullet}$ gives the desired homotopy $rg \simeq eh$ relative to $A$.

This concludes the proof of Lemma 3.4.5, and thus of Proposition 3.4.4. $\quad\square$

### 3.5. Turning Serre Fibrations into Bundles

**Proposition 3.5.1.** *(c) The functors*

$$\tilde{n} \colon s\mathcal{C} \xrightarrow{\simeq} s\widetilde{\mathcal{C}}_{\bullet}$$

*and $\tilde{n} \colon h\mathcal{C} \xrightarrow{\simeq} h\widetilde{\mathcal{C}}_{\bullet}$, given by inclusion of 0-simplices, are homotopy equivalences.*

*(d) The functors*

$$n \colon s\mathcal{D} \xrightarrow{\simeq} s\mathcal{D}_\bullet$$

*and* $n \colon h\mathcal{D} \xrightarrow{\simeq} h\mathcal{D}_\bullet$, *given by inclusion of* 0-*simplices, and the full embeddings*

$$v \colon s\mathcal{D}_\bullet \xrightarrow{\simeq} s\widetilde{\mathcal{D}}_\bullet$$

*and* $v \colon h\mathcal{D}_\bullet \xrightarrow{\simeq} h\widetilde{\mathcal{D}}_\bullet$, *are homotopy equivalences.*

In fact, these are degreewise homotopy equivalences, if we interpret $s\mathcal{C}$, $h\mathcal{C}$, $s\mathcal{D}$ and $h\mathcal{D}$ as constant simplicial categories.

*Proof.* We prove that the functors $\tilde{n} \colon s\mathcal{C} \to s\widetilde{\mathcal{C}}_q$, $\tilde{n} = vn \colon s\mathcal{D} \to s\widetilde{\mathcal{D}}_q$ and $v \colon s\mathcal{D}_q \to s\widetilde{\mathcal{D}}_q$ are homotopy equivalences, and likewise for the $h$-prefixed cases. The proposition then follows by the realization lemma, and the 2-out-of-3 property.

To each Serre fibration $\pi \colon Z \to \Delta^q$ of finite simplicial sets, we can associate the preimage of the barycenter

$$\Phi(Z) = Sd(\pi)^{-1}(\beta) \,.$$

Then $\Phi$ defines functors $s\widetilde{\mathcal{C}}_q \to s\mathcal{C}$, $s\widetilde{\mathcal{D}}_q \to s\mathcal{D}$, $h\widetilde{\mathcal{C}}_q \to h\mathcal{C}$ and $h\widetilde{\mathcal{D}}_q \to h\mathcal{D}$. For if $Z$ is non-singular, then so is $Sd(Z) \cong B(Z)$, and $\Phi(Z)$ is a simplicial subset of $Sd(Z)$, hence is non-singular. If $f \colon Z \to Z'$ over $\Delta^q$ is a simple map, then $Sd(f) \colon Sd(Z) \to Sd(Z')$ is simple by Proposition 2.3.3 (only the non-singular part is needed for case (d)), so $\Phi(f) \colon \Phi(Z) \to \Phi(Z')$ is simple by the pullback property. If $f \colon Z \to Z'$ over $\Delta^q$ is (just) a weak homotopy equivalence, then we use the commutative square

$$\begin{array}{ccc}
\Phi(Z) \times \Delta^q & \xrightarrow{\ t\ } & Z \\[2pt]
{\scriptstyle \Phi(f) \times id} \downarrow & & \downarrow {\scriptstyle f} \\[2pt]
\Phi(Z') \times \Delta^q & \xrightarrow{\ t'\ } & Z' \,,
\end{array}$$

where $t$ and $t'$ are simple by Proposition 2.7.6, and the 2-out-of-3 property, to deduce that $\Phi(f)$ is also a weak homotopy equivalence.

We claim that $\Phi$ is homotopy inverse to $\tilde{n}$, in each of the four cases. The natural simple map

$$t \colon \Phi(Z) \times \Delta^q \to Z$$

over $\Delta^q$ defines a natural transformation from $\tilde{n}\Phi$ to the identity. It does not simultaneously define a natural transformation from $\Phi\tilde{n}$ to $id$, because $\tilde{n}$ is not a full embedding. Instead, fix a morphism $\eta \colon [0] \to [q]$ in $\Delta$. For any finite simplicial set $X$ we can restrict the simple map

$$t \colon \Phi(X \times \Delta^q) \times \Delta^q \to X \times \Delta^q$$

(associated to $Z = X \times \Delta^q$) along $id \times \eta \colon X \to X \times \Delta^q$. The resulting simple map

$$\eta^* t \colon \Phi(X \times \Delta^q) \to X$$

then defines a natural transformation from $\Phi \tilde{n}$ to the identity.

The functor $v \colon s\mathcal{D}_q \to s\widetilde{\mathcal{D}}_q$ is a full embedding, so this case is easier. For $\pi \colon Z \to \Delta^q$ a Serre fibration of finite non-singular simplicial sets, let $P(Z) = \Phi(Z) \times \Delta^q$ ($P$ for product). Then $pr \colon P(Z) \to \Delta^q$ is certainly a PL bundle after geometric realization, so this defines a functor $P \colon s\widetilde{\mathcal{D}}_q \to s\mathcal{D}_q$. The simple map

$$t \colon P(Z) \to Z$$

over $\Delta^q$ then defines the two required natural transformations from $vP$ and $Pv$ to the respective identity functors. The proof for $v \colon h\mathcal{D}_q \to h\widetilde{\mathcal{D}}_q$ is the same. $\square$

**Corollary 3.5.2.** *The inclusions of $0$-simplices*

$$\tilde{n} \colon s\mathcal{C}^h(X) \to s\widetilde{\mathcal{C}}^h_\bullet(X)$$

*and*

$$\tilde{n} \colon s\mathcal{D}^h(X) \to s\widetilde{\mathcal{D}}^h_\bullet(X)$$

*are homotopy equivalences.*

*Proof.* This is clear by Proposition 3.3.1 and part of Proposition 3.5.1. See also diagram (3.5.4). $\square$

**Corollary 3.5.3.** *The full embeddings*

$$v \colon s\mathcal{E}_\bullet \to s\widetilde{\mathcal{E}}_\bullet$$

*and $v \colon h\mathcal{E}_\bullet \to h\widetilde{\mathcal{E}}_\bullet$ are homotopy equivalences.*

*Proof.* Combine Proposition 3.4.4 with the remainder of Proposition 3.5.1. $\square$

We can now prove the part of the stable parametrized $h$-cobordism theorem that compares non-singular simplicial sets to PL Serre fibrations of polyhedra.

*Proof of Theorem 1.2.6.* There is a commutative diagram

$$(3.5.4)$$

$$
\begin{array}{ccccc}
s\mathcal{D}^h(X) & \longrightarrow & s\mathcal{D} & \overset{j}{\longrightarrow} & h\mathcal{D} \\
\tilde{n}\downarrow & & \downarrow\tilde{n} & & \downarrow\tilde{n} \\
s\widetilde{\mathcal{D}}^h_\bullet(X) & \longrightarrow & s\widetilde{\mathcal{D}}_\bullet & \overset{j}{\longrightarrow} & h\widetilde{\mathcal{D}}_\bullet \\
r\downarrow & & \downarrow r & & \downarrow r \\
s\widetilde{\mathcal{E}}^h_\bullet(|X|) & \longrightarrow & s\widetilde{\mathcal{E}}_\bullet & \overset{j}{\longrightarrow} & h\widetilde{\mathcal{E}}_\bullet
\end{array}
$$

where the rows are homotopy fiber sequences (based at the points represented by $X$, $X$ and $|X|$, respectively) by Proposition 3.3.1, and $\tilde{n} = vn$. In view of the commutative diagram (3.1.8), the functor $\tilde{n} \circ r \colon s\mathcal{D}^h(X) \to s\widetilde{\mathcal{E}}^h_\bullet(|X|)$ of Theorem 1.2.6 equals the composite functor $r \circ \tilde{n}$ at the left hand side. By Proposition 3.5.1(d) the functors $\tilde{n} \colon s\mathcal{D} \to s\widetilde{\mathcal{D}}_\bullet$ and $\tilde{n} \colon h\mathcal{D} \to h\widetilde{\mathcal{D}}_\bullet$ (inclusion of 0-simplices) are homotopy equivalences. By Proposition 3.4.4 the functors $r \colon s\widetilde{\mathcal{D}}_\bullet \to s\widetilde{\mathcal{E}}_\bullet$ and $r \colon h\widetilde{\mathcal{D}}_\bullet \to h\widetilde{\mathcal{E}}_\bullet$ (polyhedral realization) are homotopy equivalences. Hence the middle and right hand composite functors $r \circ \tilde{n}$ in (3.5.4) are homotopy equivalences. By the homotopy fiber property, it follows that also the left hand composite $r \circ \tilde{n}$ is a homotopy equivalence.  $\square$

*Remark 3.5.5.* Case (d) of Proposition 3.3.1, and thus of Lemma 3.2.9, was not really needed for the above proof, because we did not need to know that the middle row is a homotopy fiber sequence. That the top row is a homotopy fiber sequence follows from Propositions 3.3.1(c) and 3.1.14, see diagram (3.1.15). The claim for the bottom row follows from Proposition 3.3.1(e).

## 3.6. QUILLEN'S THEOREMS A AND B

For the reader's convenience, we recall Quillen's Theorems A and B from [Qu73, §1] and their variants for simplicial categories, Theorems A′ and B′ from [Wa82, §4]. We state the versions for right fibers. In each case there is also a dual formulation in terms of left fibers. We also recall the realization lemma from [Se74, App. A].

**Definition 3.6.1.** Let $\mathcal{A}$ and $\mathcal{B}$ be categories and $f \colon \mathcal{A} \to \mathcal{B}$ a functor. For each object $Y$ of $\mathcal{B}$, let the **right fiber** $Y/f$ of $f$ at $Y$ be the category of pairs $(X, b)$ where $X$ is an object of $\mathcal{A}$ and $b \colon Y \to f(X)$ is a morphism in $\mathcal{B}$. A morphism in $Y/f$ from $(X, b)$ to $(X', b')$ is a morphism $a \colon X \to X'$ in $\mathcal{A}$ such that $b' = f(a) \circ b$.

$$
\begin{array}{ccc}
X & \xrightarrow{\ a\ } & X' \\
{\scriptstyle f}\big\downarrow & & \big\downarrow{\scriptstyle f} \\
Y \xrightarrow{\ b\ } f(X) & \xrightarrow{f(a)} & f(X')
\end{array}
$$

Dually, let the **left fiber** $f/Y$ be the category of pairs $(X, b)$ where $X$ is an object of $\mathcal{A}$ and $b \colon f(X) \to Y$ is a morphism in $\mathcal{B}$. A morphism in $f/Y$ from $(X, b)$ to $(X', b')$ is a morphism $a \colon X \to X'$ in $\mathcal{A}$ such that $b = b' \circ f(a)$.

**Theorem 3.6.2 (Quillen's Theorem A).** *Let* $f \colon \mathcal{A} \to \mathcal{B}$ *be a functor. If the right fiber* $Y/f$ *is contractible for every object* $Y$ *of* $\mathcal{B}$, *then* $f$ *is a homotopy equivalence.*

**Definition 3.6.3.** For each morphism $\beta \colon Y \to Y'$ in $\mathcal{B}$, let the **transition**

**map** $\beta^*\colon Y'/f \to Y/f$ be the functor that takes $(X', b')$ to $(X', b' \circ \beta)$.

$$
\begin{array}{c}
X' \\
\downarrow f \\
Y \xrightarrow{\ \beta\ } Y' \xrightarrow{\ b'\ } f(X)
\end{array}
$$

Dually, let the transition map $\beta_*\colon f/Y \to f/Y'$ of left fibers be the functor that takes $(X, b)$ to $(X, \beta \circ b)$.

There is a canonical functor $Y/f \to \mathcal{A}$ that takes $(X, b)$ to $X$. We write $Y/\mathcal{B}$ for the right fiber of $id_{\mathcal{B}}$ at $Y$. This category has the initial object $(Y, id_Y)$, hence is contractible. There is a functor $Y/f \to Y/\mathcal{B}$ that takes $(X, b)$ to $(f(X), b)$, and the canonical functor $Y/\mathcal{B} \to \mathcal{B}$ takes $(Y', b\colon Y \to Y')$ to $Y'$. Together with $f$, these functors form the commutative square in the following result.

**Theorem 3.6.4 (Quillen's Theorem B).** *Let $f\colon \mathcal{A} \to \mathcal{B}$ be a functor. If the transition map $\beta^*\colon Y'/f \to Y/f$ is a homotopy equivalence for every morphism $\beta\colon Y \to Y'$ in $\mathcal{B}$, then for any object $Y$ of $\mathcal{B}$ the commutative square*

$$
\begin{array}{ccc}
Y/f & \longrightarrow & \mathcal{A} \\
\downarrow & & \downarrow f \\
Y/\mathcal{B} & \longrightarrow & \mathcal{B}
\end{array}
$$

*is homotopy cartesian. The term $Y/\mathcal{B}$ is contractible.*

Quillen's Theorems A and B are proved in [Qu73, pp. 93–99].

*Remark 3.6.5.* The conclusions of these theorems (to the eyes of homology with local coefficients) can be deduced from the spectral sequence of Theorem 3.6 in [GZ67, App. II], using Paragraph 3.5 and Remark 3.8 of that appendix, respectively.

**Definition 3.6.6.** Let $\mathcal{A}_\bullet$ and $\mathcal{B}_\bullet$ be simplicial categories and $f\colon \mathcal{A}_\bullet \to \mathcal{B}_\bullet$ a simplicial functor. By an object of $\mathcal{B}_\bullet$ we mean a pair $([q], Y)$, where $q \geq 0$ and $Y$ is an object of $\mathcal{B}_q$. For each object $([q], Y)$ of $\mathcal{B}_\bullet$ let the **right fiber**

$$([q], Y)/f$$

of $f$ at $([q], Y)$ be the simplicial category

$$[n] \mapsto \coprod_{\varphi\colon [n] \to [q]} \varphi^* Y / f_n .$$

Here the coproduct (disjoint union of categories) is indexed by the set of morphisms $\varphi\colon [n] \to [q]$ in $\Delta$, and $\varphi^* Y / f_n$ is the right fiber at $\varphi^* Y$ of the functor $f_n\colon \mathcal{A}_n \to \mathcal{B}_n$.

Dually, in degree $n$ the **left fiber** $f/([q], Y)$ is the coproduct over all $\varphi\colon [n] \to [q]$ of the left fibers $f_n/\varphi^* Y$.

**Theorem 3.6.7 (Theorem A').** *Let* $f\colon \mathcal{A}_\bullet \to \mathcal{B}_\bullet$ *be a simplicial functor. If the right fiber* $([q], Y)/f$ *is contractible for every object* $([q], Y)$ *of* $\mathcal{B}_\bullet$, *then* $f$ *is a homotopy equivalence.*

**Definition 3.6.8.** For each morphism $\beta\colon Y \to Y'$ in $\mathcal{B}_q$, in degree $q$, let the **transition map (of the first kind)** be the simplicial functor

$$([q], \beta)^*\colon ([q], Y')/f \to ([q], Y)/f$$

given in degree $n$ by the coproduct over all morphisms $\varphi\colon [n] \to [q]$ in $\Delta$ of the transition maps

$$(\varphi^*\beta)^*\colon \varphi^*Y'/f_n \to \varphi^*Y/f_n.$$

Here $\varphi^*\beta\colon \varphi^*Y \to \varphi^*Y'$ in $\mathcal{B}_n$.

For each morphism $\alpha\colon [p] \to [q]$ in $\Delta$, let the **transition map (of the second kind)** be the simplicial functor

$$\alpha_*\colon ([p], \alpha^*Y)/f \to ([q], Y)/f$$

that in degree $n$ takes the $\varphi$-summand to the $\alpha\varphi$-summand by the identity functor

$$id\colon \varphi^*\alpha^*Y/f_n \to (\alpha\varphi)^*Y/f_n,$$

for each morphism $\varphi\colon [n] \to [p]$ in $\Delta$.

Dually, there are transition maps $([q], \beta)_*\colon f/([q], Y) \to f/([q], Y')$ (of the first kind) and $\alpha_*\colon f/([p], \alpha^*Y) \to f/([q], Y)$ (of the second kind) of left fibers.

We write $([q], Y)/\mathcal{B}_\bullet$ for the right fiber of $id_{\mathcal{B}_\bullet}$ at the object $([q], Y)$ in $\mathcal{B}_\bullet$. In degree $n$ it is the coproduct of the categories $\varphi^*Y/\mathcal{B}_n$, over all morphisms $\varphi\colon [n] \to [q]$, each of which has an initial object and is therefore contractible. Hence $([q], Y)/\mathcal{B}_\bullet$ maps by a homotopy equivalence to the $q$-simplex $\Delta^q$, considered as a simplicial category in a trivial way, and is therefore contractible.

There is a canonical simplicial functor $([q], Y)/f \to \mathcal{A}_\bullet$, induced in degree $n$ from the functors $\varphi^*Y/f_n \to \mathcal{A}_n$, and similarly for $([q], Y)/\mathcal{B}_\bullet \to \mathcal{B}_\bullet$. There is also a simplicial functor $([q], Y)/f \to ([q], Y)/\mathcal{B}_\bullet$ that in degree $n$ is the coproduct of the usual functors $\varphi^*Y/f_n \to \varphi^*Y/\mathcal{B}_n$.

**Theorem 3.6.9 (Theorem B').** *Let* $f\colon \mathcal{A}_\bullet \to \mathcal{B}_\bullet$ *be a simplicial functor. If all transition maps (of the first and the second kind) of right fibers are homotopy equivalences, then for any object* $([q], Y)$ *in* $\mathcal{B}_\bullet$ *the commutative square*

$$
\begin{array}{ccc}
([q], Y)/f & \longrightarrow & \mathcal{A}_\bullet \\
\downarrow & & \downarrow{\scriptstyle f} \\
([q], Y)/\mathcal{B}_\bullet & \longrightarrow & \mathcal{B}_\bullet
\end{array}
$$

*is homotopy cartesian.*

Theorems A$'$ and B$'$ are deduced from Quillen's Theorems A and B in [Wa82, §4], using [Th79] and [Se74].

We also recall from [Wa82, §4] the following useful addendum. An object $Y$ of $\mathcal{B}_q$ will be said to be 0-**dimensional up to isomorphism** if there exists an object $Z$ of $\mathcal{B}_0$ and an isomorphism $Y \cong \epsilon^* Z$ in $\mathcal{B}_q$, where $\epsilon\colon [q] \to [0]$ is the unique morphism in $\Delta$.

**Addendum 3.6.10.** *If every object of $\mathcal{B}_\bullet$ is 0-dimensional up to isomorphism, then the hypothesis of Theorem A$'$, resp. of Theorem B$'$, needs only be checked in the case $q = 0$.*

*Proof.* Suppose that $Y = \epsilon^* Z$ in $\mathcal{B}_q$, with $Z$ in $\mathcal{B}_0$ and $\epsilon\colon [q] \to [0]$. Then $\varphi^* Y = (\epsilon\varphi)^* Z$ is independent of $\varphi\colon [n] \to [q]$, so

$$([q], Y)/f \cong \Delta^q \times ([0], Z)/f$$

and the transition map $\epsilon_*\colon ([q], Y)/f \to ([0], Z)/f$ is a homotopy equivalence. It follows that if every object of $\mathcal{B}$ is isomorphic to one of the form $\epsilon^* Z$, with $Z$ in degree 0, then every right fiber of $f$ is homotopy equivalent to one of the form $([0], Z)/f$.

For $Y = \epsilon^* Z$ as above, and any morphism $\alpha\colon [p] \to [q]$ in $\Delta$, we have just seen that $\epsilon_* \alpha_* = (\epsilon\alpha)_*$ and $\epsilon_*$ are homotopy equivalences, so $\alpha_*\colon ([p], \alpha^* Y)/f \to ([q], Y)/f$ is a homotopy equivalence by the 2-out-of-3 property. It follows that if every object of $\mathcal{B}$ is 0-dimensional up to isomorphism, then every transition map of the second kind is a homotopy equivalence.

Let $\beta\colon Y \to Y'$ be any morphism in $\mathcal{B}_q$, and let $\eta\colon [0] \to [q]$ be any morphism in $\Delta$. We have a commutative square

$$
\begin{array}{ccc}
([0], \eta^* Y')/f & \xrightarrow{\;([0], \eta^* \beta)^*\;} & ([0], \eta^* Y)/f \\
\Big\downarrow{\scriptstyle \eta_*} & & \Big\downarrow{\scriptstyle \eta_*} \\
([q], Y')/f & \xrightarrow{\;([q], \beta)^*\;} & ([q], Y)/f\,.
\end{array}
$$

Still assuming that every object of the target simplicial category $\mathcal{B}_\bullet$ is 0-dimensional up to isomorphism, we have just seen that the vertical transition maps $\eta_*$ are homotopy equivalences. Hence every transition map of the first kind is homotopy equivalent to one of the form $([0], \beta')^*$, where $\beta'$ is a morphism in $\mathcal{B}_0$. $\square$

The expression for the right fiber $([0], Z)/f$ simplifies to the simplicial category

$$[q] \mapsto \epsilon^* Z / f_q\,.$$

Elsewhere in this book, when the addendum applies, we sometimes simply write $Z/f$ for $([0], Z)/f$. Note that the objects of the source simplicial category $\mathcal{A}_\bullet$ are not required to be 0-dimensional up to isomorphism. Our Lemmas 3.3.6 and 3.3.7 exhibit a more subtle situation in which one can reduce to checking the hypotheses of Theorem B' in simplicial degree 0, even if Addendum 3.6.10 does not apply.

Recall that a bisimplicial set

$$Z_{\bullet,\bullet}\colon [p], [q] \mapsto Z_{p,q}$$

can be viewed as a simplicial object $X_\bullet\colon [p] \mapsto X_p$ in the category of simplicial sets, where $X_p$ is the simplicial set $Z_{p,\bullet}\colon [q] \mapsto Z_{p,q}$, for each $p \geq 0$. The geometric realization of $X_\bullet$ equals the geometric realization of $Z_{\bullet,\bullet}$, and is isomorphic to the geometric realization of the diagonal simplicial set $diag(Z)_\bullet\colon [p] \mapsto Z_{p,p}$.

The following result is known as the "realization lemma."

**Proposition 3.6.11.** *Let $f\colon X_\bullet \to Y_\bullet$ be a map of bisimplicial sets, viewed as simplicial objects in the category of simplicial sets. If each map $f_p\colon X_p \to Y_p$ is a weak homotopy equivalence, for $p \geq 0$, then $f$ is a weak homotopy equivalence.*

*Proof.* This is effectively proved in [Se74, App. A]. The simplicial space $[p] \mapsto |X_p|$ is good in the sense of [Se74, Def. A.4], because each degeneracy operator $[p] \to [n]$ admits a section, hence induces a split injection $X_n \to X_p$ of simplicial sets, and thus induces a closed cofibration $|X_n| \to |X_p|$ upon geometric realization. The same applies to $[p] \mapsto |Y_p|$, so the result follows from [Se74, Prop. A.1(ii) and (iv)], which establishes the following commutative diagram.

$$
\begin{array}{ccc}
\|X_\bullet\| & \xrightarrow[\simeq]{\|f\|} & \|Y_\bullet\| \\
{\scriptstyle\simeq}\downarrow & & \downarrow{\scriptstyle\simeq} \\
|X_\bullet| & \xrightarrow{|f|} & |Y_\bullet|
\end{array}
$$

A key step in Segal's proof of Proposition A.1(iv) uses a union theorem (or gluing lemma) for closed cofibrations, which in turn is proved in [Li73]. $\square$

# Chapter Four

# The manifold part

In this chapter, let $\Delta^q$ denote the standard affine $q$-simplex. All polyhedra will be compact, and all manifolds considered will be compact PL manifolds, usually without further mention. Recall the fixed intervals $I$ and $J$ from Section 1.1.

## 4.1. Spaces of PL manifolds

The aim of this section is to reduce the proof of the manifold part of the stable parametrized $h$-cobordism theorem, Theorem 1.1.8, to a result about spaces of stably framed manifolds, Theorem 4.1.14, which will be proved in the following two sections.

**Definition 4.1.1.** Let $P$ be a compact polyhedron. By a **family of manifolds parametrized by** $P$ we shall mean a **PL bundle** (= a PL locally trivial family) $\pi\colon E \to P$ whose fibers $M_p = \pi^{-1}(p)$ are compact PL manifolds. A map of such families is a PL bundle map over $P$.

We shall define a space of manifolds as a simplicial set, with families of manifolds parametrized by $\Delta^q$ as the $q$-simplices. To properly compare this space with a space of polyhedra, we might either equip the polyhedra with tangential information, or else arrange that the manifolds have no tangent bundle, that is, to work with framed manifolds. Another way, which is equivalent to the latter one in the stable range that we will consider, is to arrange that all the manifolds are embedded with codimension zero in Euclidean space. Which of these alternatives to choose is a matter of convenience — we have chosen to work with stable framings.

We rely on [Mi64], [HP64], [HW65] and [KL66] for the theory of PL microbundles. The **tangent microbundle** $\tau_M$ of a PL manifold $M$ is (the equivalence class of) the diagram

$$M \xrightarrow{\Delta} M \times M \xrightarrow{pr} M$$

where $\Delta(x) = (x, x)$ and $pr(x, y) = x$. If $\pi\colon E \to P$ is a family of PL manifolds, then the tangent microbundles $\tau_{M_p}$ of the fiber manifolds $M_p = \pi^{-1}(p)$ can be combined to a **fiberwise tangent microbundle** $\tau_E^\pi$, given by the diagram

$$E \xrightarrow{\Delta} E \times_P E \xrightarrow{pr} E,$$

where $E \times_P E = \{(e, f) \in E \times E \mid \pi(e) = \pi(f)\}$ is the fiber product, $\Delta(e) = (e, e)$ and $pr(e, f) = e$.

**Definition 4.1.2.** Let $M$ be a manifold of dimension $n$. A **stable framing** of $M$ is an equivalence class of isomorphisms

$$\tau_M \oplus \epsilon^k \xrightarrow{\cong} \epsilon^{n+k}$$

of PL microbundles, where $\tau_M$ is the tangent microbundle of $M$ and $\epsilon^k$ is a standard trivial $k$-bundle. The equivalence relation is generated by the rule that the isomorphism above may be replaced with the isomorphism

$$\tau_M \oplus \epsilon^{k+1} \xrightarrow{\cong} \epsilon^{n+k+1}$$

obtained by Whitney sum with $\epsilon^1$.

By a **stably framed family of manifolds** parametrized by $P$ we shall mean a family of manifolds $\pi \colon E \to P$ such that for each $p \in P$ the fiber manifold $M_p = \pi^{-1}(p)$ is stably framed, and these stable framings vary in a PL manner with the point $p$. In other words, the PL microbundle isomorphisms $\tau_{M_p} \oplus \epsilon^k \xrightarrow{\cong} \epsilon^{n+k}$ are required to combine to a PL microbundle isomorphism

$$\tau_E^\pi \oplus \epsilon^k \xrightarrow{\cong} \epsilon^{n+k} .$$

A codimension zero PL embedding $f \colon M \to N$, or more generally, a codimension zero PL immersion, induces a PL microbundle isomorphism $\tau_M \cong f^* \tau_N$, so a stable framing of $N$ pulls back along $f$ to determine a stable framing of $M$. We say that such a map $f$ is **stably framed** if $M$ and $N$ are stably framed and the given stable framing of $N$ pulls back along $f$ to the given stable framing of $M$.

Note that a stable framing of a manifold $M$ can induce one of its boundary $\partial M$, even in parametrized families. To justify this, one uses the contractibility of the space of collars on the boundary (see Remark 1.1.2).

**Definition 4.1.3.** The **space of stably framed $n$-manifolds** $\mathcal{M}^n_\bullet$ is defined as the simplicial set whose $q$-simplices are the stably framed families of $n$-manifolds parametrized by $\Delta^q$. Similarly, $h\mathcal{M}^n_\bullet$ is defined as the simplicial category whose objects in simplicial degree $q$ are the same as the $q$-simplices of $\mathcal{M}^n_\bullet$, and whose morphisms in simplicial degree $q$ are the PL bundle maps over $\Delta^q$ that are also homotopy equivalences.

There are stabilization maps $\sigma \colon \mathcal{M}^n_\bullet \to \mathcal{M}^{n+1}_\bullet$ and $\sigma \colon h\mathcal{M}^n_\bullet \to h\mathcal{M}^{n+1}_\bullet$, defined by multiplying each manifold bundle with the fixed closed interval $J$. There is a natural inclusion $j \colon \mathcal{M}^n_\bullet \to h\mathcal{M}^n_\bullet$, which commutes with the stabilization maps.

To ensure that the collection $\mathcal{M}^n_q$ of $q$-simplices is really a set, we proceed as in Remark 1.1.2, and assume that each bundle $E \to \Delta^q$ is embedded in $\mathbb{R}^\infty \times \Delta^q \to \Delta^q$. The simplicial operators are also defined as in that remark, via pullback and a canonical identification.

Recall the PL $h$-cobordism space $H^{PL}(M)$ and its stabilization $\mathcal{H}^{PL}(M)$, from Definitions 1.1.1 and 1.1.3, together with their collared versions $H^{PL}(M)^c$ and $\mathcal{H}^{PL}(M)^c$. The forgetful maps $H^{PL}(M)^c \to H^{PL}(M)$ and $\mathcal{H}^{PL}(M)^c \to \mathcal{H}^{PL}(M)$ are weak homotopy equivalences. To get a natural map from $H^{PL}(M)$ to $\mathcal{M}_\bullet^n$, for $n = \dim(M \times I)$, it will be convenient to introduce a third model for the $h$-cobordism space, weakly equivalent to the other two, where we have also chosen a stable framing of the collared $h$-cobordisms.

**Definition 4.1.4.** For a stably framed manifold $M$, let $H(M)^f = H^{PL}(M)^f$ be the space of **collared, stably framed** PL $h$-cobordisms on $M$. A 0-simplex is a PL $h$-cobordism $W$ on $M$, with a collar $c\colon M \times I \to W$, together with a stable framing of $W$ that restricts via $c$ to the product stable framing of $M \times I$. A $q$-simplex is a PL bundle of stably framed $h$-cobordisms over $\Delta^q$, relative to the embedded product bundle $pr\colon M \times I \times \Delta^q \to \Delta^q$.

Each stably framed codimension zero embedding $M \to M'$ induces a map $H(M)^f \to H(M')^f$ that takes $W$ to the $h$-cobordism $W' = M' \times I \cup_{M \times I} W$, with the stable framing that restricts to the product stable framing of $M' \times I$ and the given stable framing of $W$.

There is a natural stabilization map $\sigma\colon H(M)^f \to H(M \times J)^f$, which takes the stable framing of $W$ to the product stable framing of $W \times J$. Since $c$ is a homotopy equivalence, the natural forgetful maps $H(M)^f \to H(M)^c$ and

$$\mathcal{H}^{PL}(M)^f = \operatorname*{colim}_k H^{PL}(M \times J^k)^f \to \mathcal{H}^{PL}(M)^c$$

are both weak homotopy equivalences.

The following result is similar to the approximate homotopy fiber sequence

$$H(M) \to \mathcal{P}_0^m(M) \to h\mathcal{P}_0^m(M)$$

from [Wa82, Prop. 5.1], valid for a range of homotopy groups that grows to infinity with the dimension of $M$. This is the case $k = 0$ (no handles), so the handle index $m$ plays no role. See also [Wa87b, Exer.], in connection with Proposition 4.1.15.

**Proposition 4.1.5.** *Let $M$ be a stably framed compact PL manifold. There is a natural homotopy fiber sequence*

$$\mathcal{H}^{PL}(M)^f \to \operatorname*{colim}_n \mathcal{M}_\bullet^n \xrightarrow{j} \operatorname*{colim}_n h\mathcal{M}_\bullet^n$$

*based at $M$.*

*Proof.* The argument has two parts. First, we shall use the simplicial form, Theorem B$'$, of Quillen's Theorem B to prove that there is a natural homotopy fiber sequence

$$M/j \to \operatorname*{colim}_n \mathcal{M}_\bullet^n \xrightarrow{j} \operatorname*{colim}_n h\mathcal{M}_\bullet^n$$

based at $M$. Thereafter, in the following Proposition 4.1.6, we will show that there is a natural weak homotopy equivalence

$$\mathcal{H}^{PL}(M)^f \xrightarrow{\simeq} M/j\,.$$

More precisely, we shall prove these results with $\mathrm{colim}_k(M \times I \times J^k)/j$ in place of $M/j$. When combined, these two results complete the proof.

We begin by verifying that Theorem 3.6.9 (= Theorem B$'$) and Addendum 3.6.10 apply to the unstabilized functor

$$j\colon \mathcal{M}^n_\bullet \to h\mathcal{M}^n_\bullet\,.$$

The addendum applies because each PL bundle $\pi\colon E \to \Delta^q$ of stably framed $n$-manifolds is isomorphic to a product bundle of the same kind.

Let $N$ be any stably framed $n$-manifold, viewed as an object of $h\mathcal{M}^n_0$. The right fiber $N/j = ([0], N)/j$ is the simplicial set of stably framed $n$-manifolds $V$ with a PL homotopy equivalence $g\colon N \to V$, and similarly in parametrized families. So a $q$-simplex of $N/j$ is a PL bundle map and homotopy equivalence $e\colon N \times \Delta^q \to E$ of stably framed manifold families over $\Delta^q$.

Let $\beta\colon N \to N'$ be any morphism in $h\mathcal{M}^n_0$, i.e., a PL homotopy equivalence. Let $\beta'\colon N' \to N$ be a PL homotopy inverse map. To show that the transition map $\beta^*\colon N'/j \to N/j$ is a homotopy equivalence, it suffices to show that the transition maps $(\beta'\beta)^*$ and $(\beta\beta')^*$ are homotopy equivalences. Hence it suffices to consider the case $N = N'$, and to show that any PL homotopy $\beta \simeq id\colon N \to N$ induces a homotopy of transition maps $\beta^* \simeq id\colon N/j \to N/j$.

Let $H\colon \Delta^1 \times N \to N$ be a PL homotopy from $\beta$ to $id$. Then there is a simplicial homotopy

$$H^*\colon \Delta^1 \times N/j \to N/j$$

from $\beta^*$ to $id$ that to each morphism $\alpha\colon [q] \to [1]$ in $\Delta$ and $q$-simplex $e\colon N \times \Delta^q \to E$ of $N/j$ associates the $q$-simplex of $N/j$ given by pullback along $(\alpha, id)\colon \Delta^q \to \Delta^1 \times \Delta^q$ of the composite PL bundle map

$$\Delta^1 \times N \times \Delta^q \xrightarrow{(pr,H)\times id} \Delta^1 \times N \times \Delta^q \xrightarrow{id \times e} \Delta^1 \times E$$

over $\Delta^1 \times \Delta^q$. Cf. [Wa85, p. 335] for more on this way of describing simplicial homotopies. Each of $(pr, H) \times id$ and $id \times e$ is a fiberwise homotopy equivalence, so the pullback over $\Delta^q$ is also a homotopy equivalence, and defines a $q$-simplex of $N/j$.

Thus, by Theorem B$'$ there is a homotopy fiber sequence $N/j \to \mathcal{M}^n_\bullet \xrightarrow{j} h\mathcal{M}^n_\bullet$ based at $N$. Letting $N$ range through the stably framed manifolds $M \times I \times J^k$, for $k \geq 0$, the resulting homotopy fiber sequences are compatible under the stabilization maps. Their colimit is a homotopy fiber sequence

$$\operatorname*{colim}_k (M \times I \times J^k)/j \to \operatorname*{colim}_n \mathcal{M}^n_\bullet \xrightarrow{\mathrm{colim}_n j} \operatorname*{colim}_n h\mathcal{M}^n_\bullet$$

based at $M$. Abbreviating the stabilized functor $\mathrm{colim}_n\, j$ to $j$, we have the asserted homotopy fiber sequence. $\square$

**Proposition 4.1.6.** *Let*

$$\varphi^f \colon H(M \times J^k)^f \to (M \times I \times J^k)/j$$

*be the natural map that to a collared, stably framed $h$-cobordism $(W, c)$ on $M \times J^k$ associates the same pair $(W, c)$, where now $W$ is viewed as a stably framed manifold, and $c \colon M \times I \times J^k \to W$ is viewed as a PL homotopy equivalence. Then the stabilized map*

$$\operatorname*{colim}_k \varphi^f \colon \mathcal{H}^{PL}(M)^f \to \operatorname*{colim}_k (M \times I \times J^k)/j$$

*is a weak homotopy equivalence.*

*Proof.* For brevity, we will write $N = M \times I \times J^k$, with the stable framing obtained by stabilization from the given one of $M$. Let $d = \dim(M)$ and $n = \dim(N)$, so $n = d + 1 + k$.

We have a commutative square

$$
\begin{array}{ccc}
H(M \times J^k)^f & \xrightarrow{\;\varphi^f\;} & N/j = \{g \colon N \xrightarrow{\simeq} V,\ \tau_V \text{ stably trivialized}\} \\[2pt]
{\scriptstyle\simeq}\Big\downarrow & & \Big\downarrow{\scriptstyle\simeq} \\[4pt]
H(M \times J^k)^c & \xrightarrow{\;\varphi\;} & \{g \colon N \xrightarrow{\simeq} V,\ g^*\tau_V \text{ stably trivialized}\}\,.
\end{array}
$$

The lower right hand space is the simplicial set with 0-simplices consisting of $n$-manifolds $V$ with a *PL* homotopy equivalence $g \colon N \to V$ and a stable trivialization of the pullback $g^*\tau_V$ of the tangent microbundle of $V$. Note that $N$ and $V$ have the same dimension. The $q$-simplices consist of PL bundles over $\Delta^q$ of the same kind of data.

The upper horizontal map $\varphi^f$ was described above, the left hand vertical map is the weak homotopy equivalence that forgets the stable framing, and the right hand vertical map pulls the stable trivialization of $\tau_V$ back along $g$ to give a stable trivialization of $g^*\tau_V$. It is a weak homotopy equivalence because $g$ is a homotopy equivalence, and stable trivializations are given in terms of PL microbundle data. The lower horizontal map $\varphi$ takes a collared $h$-cobordism $(W, c)$ on $M \times J^k$ to $(V, g) = (W, c)$, with the stable trivialization of $c^*\tau_W$ that is determined by the fixed stable trivialization of $\tau_N$ and the PL microbundle isomorphism $\hat{c} \colon \tau_N \cong c^*\tau_W$ induced by the PL embedding $c \colon N \to W$.

To complete the proof, we must prove that $\operatorname{colim}_k \varphi$ is a weak equivalence. Without loss of generality, we may and will assume that $M$ is connected, and that $k \geq 3$. Then we can factor $\varphi$ as a composite of five maps, each of which is a stable equivalence in the sense that it becomes a weak equivalence upon passage to the colimit over $k$.

$$H(M \times J^k)^c \xrightarrow{(1)} \{PL \text{ embeddings } f\colon N \xrightarrow{\simeq} V,\ \pi_1\text{-iso. } \partial V \subset V\}$$

$$\xrightarrow{(2)} \{PL \text{ embeddings } f\colon N \xrightarrow{\simeq} V\}$$

$$\xrightarrow{(3)} \{PL \text{ immersions } f\colon N \xrightarrow{\simeq} V\}$$

$$\xrightarrow{(4)} \{g\colon N \xrightarrow{\simeq} V,\ \hat{g}\colon \tau_N \cong g^*\tau_V\}$$

$$\xrightarrow{(5)} \{g\colon N \xrightarrow{\simeq} V,\ g^*\tau_V \text{ stably trivialized}\}$$

Each space is a simplicial set, with 0-simplices as described and $q$-simplices given by PL bundles over $\Delta^q$, as before. In each case $V$ has the same dimension as $N$.

The first map

$$H(M \times J^k)^c \xrightarrow{(1)} \{PL \text{ embeddings } f\colon N \xrightarrow{\simeq} V,\ \pi_1\text{-iso. } \partial V \subset V\}$$

takes the most work. It maps a collared $h$-cobordism $c\colon N \to W$ on $M \times J^k$ to the PL embedding $f = c$, with $V = W$. To check that $\partial W \subset W$ is a $\pi_1$-isomorphism, decompose $\partial W$ as the union $(M \times J^k) \cup L$ of two codimension zero submanifolds along their common boundary $\partial(M \times J^k) = \partial L$. By hypothesis, the inclusions $M \times J^k \subset W$ and $L \subset W$ are homotopy equivalences. Since $k \geq 3$ the inclusion $\partial(M \times J^k) \subset M \times J^k$ is a $\pi_1$-isomorphism, so by van Kampen's theorem the inclusion $L \subset \partial W$ is a $\pi_1$-isomorphism. Hence $\partial W \subset W$ is a $\pi_1$-isomorphism.

We now argue that this first map is homotopic to a composition

$$(4.1.7) \quad H(M \times J^k)^c \to H(\partial N)^c \xrightarrow{\cong} \{PL \text{ emb. } f\colon N \xrightarrow{\simeq} V,\ \pi_1\text{-iso. } \partial V \subset V\},$$

where the left hand map is known to be $(k-2)$-connected and the right hand map is an isomorphism.

Let $e_1\colon N = M \times I \times J^k \to \partial N \times I$ be the product of the $(k-1)$-connected codimension zero embedding $M \times J^k \cong M \times 1 \times J^k \subset \partial N$ and the identity map. The left hand map is the induced map of collared $h$-cobordism spaces, as in Definition 1.1.3(b), taking $(W, c)$ to the collared $h$-cobordism

$$c'\colon \partial N \times I \to W' := \partial N \times I \cup_N W,$$

where the pushout is formed along the embeddings $e_1$ and $c$. By Burghelea–Lashof–Rothenberg [BLR75, Thm 3.1′] (a consequence of Morlet's disjunction lemma, see (1.1.13) for the translation from concordance spaces to $h$-cobordism spaces), this map is $(k-2)$-connected.

Let $e_2\colon \partial N \times I \to N$ be an interior collar on $\partial N \subset N$, extending the identification $\partial N \times 1 \cong \partial N$. The right hand map takes a collared $h$-cobordism $c'\colon \partial N \times I \to W'$ on $\partial N$ to the PL embedding

$$f\colon N \to V := N \cup_{\partial N \times I} W'.$$

The pushout is formed along the embeddings $e_2$ and $c'$, and $f$ is a homotopy equivalence, because $c'$ is one. The inclusion $\partial V \subset V$ is a $\pi_1$-isomorphism, because $W'$ is an $h$-cobordism and $e_2$ is a $\pi_1$-isomorphism.

The embeddings $e_1$ and $e_2$ are illustrated in Figure 1. The arrows labeled $N$ indicate the product of $M \times J^k$ and $I$. The arrows labeled $V$ indicate the whole of the figure. There are two copies of $\partial N$ visible here; it should be clear from the context which copy is intended. In the notation $H(\partial N)^c$ we always mean the inner copy of $\partial N$, which is indicated by the dashed arrows.

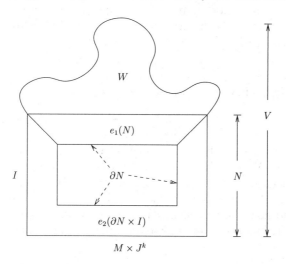

FIGURE 1. The collared $h$-cobordism $c\colon N \to W$ and the PL embedding $f\colon N \to V$, with $N = M \times I \times J^k$

Conversely, for any codimension zero PL embedding $f\colon N \to V$, the target $V$ can be decomposed as a union $N \cup_{\partial N \times I} W'$, where $\partial N \times I \to W'$ is a collar on the boundary component $\partial N \subset \partial W'$ in $W'$, and $\partial N \times I \to N$ is the same map $e_2$ as before. We claim that if $f$ is a homotopy equivalence and $\partial V \subset V$ is a $\pi_1$-isomorphism, then $W'$ is an $h$-cobordism from $\partial N$ to $\partial V$.

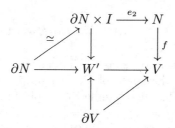

The argument is similar to that in Remark 1.1.4, and goes as follows: The embedding $W' \to V$ is a $\pi_1$-isomorphism, because $e_2$ is one. Hence the inclusion

$\partial N \subset W'$ is a $\pi_1$-isomorphism, because $f$ is one, and the inclusion $\partial V \subset W'$ is a $\pi_1$-isomorphism, because $\partial V \subset V$ is one. The assumption that $f \colon N \to V$ is a homotopy equivalence, and the excision theorem, then ensure that $\partial N \simeq \partial N \times I \to W'$ is a homology isomorphism with arbitrary local coefficients. By the universal coefficient theorem, and Lefschetz duality, it follows that $W'$ is an $h$-cobordism on $\partial N$.

This proves that the right hand map in (4.1.7) is an isomorphism. Finally, there is an isotopy from the composite embedding $e_2 e_1 \colon N \to N$ to the identity, which induces a homotopy from the composite map in (4.1.7) to the first map. Hence the map (1) is $(k-2)$-connected, and stabilizes to become a weak equivalence in the colimit.

The second map

$$\{PL \text{ embeddings } f \colon N \xrightarrow{\simeq} V, \ \pi_1\text{-iso. } \partial V \subset V\}$$
$$\xrightarrow{(2)} \{PL \text{ embeddings } f \colon N \xrightarrow{\simeq} V\}$$

forgets the condition that the inclusion $\partial V \subset V$ is a $\pi_1$-isomorphism. Given a PL embedding $f \colon N \to V$ that is a homotopy equivalence, it is not necessarily true that the inclusion $\partial V \subset V$ is a $\pi_1$-isomorphism. However, after stabilizing three times, the PL embedding $\sigma^3(f) \colon N \times J^3 \to V \times J^3$ is still a homotopy equivalence, and now the inclusion $\partial(V \times J^3) \subset V \times J^3$ is a $\pi_1$-isomorphism. Hence the map (2) induces an isomorphism upon passage to the colimit over $k$.

The third map

$$\{PL \text{ embeddings } f \colon N \xrightarrow{\simeq} V\} \xrightarrow{(3)} \{PL \text{ immersions } f \colon N \xrightarrow{\simeq} V\}$$

includes the space of codimension zero PL embeddings $f \colon N \to V$ that are homotopy equivalences into the corresponding space of codimension zero PL immersions that are homotopy equivalences. Since $N = M \times I \times J^k$, $N$ admits a handle decomposition with all handles of index $\leq \dim(M) = d$, and $V$ has dimension $n = \dim(N) = d + 1 + k$, so by general position for handle cores this inclusion is $(k-d)$-connected. Thus the connectivity of the third map grows to infinity with $k$.

The fourth map is the "differential"

$$\{PL \text{ immersions } f \colon N \xrightarrow{\simeq} V\} \xrightarrow{(4)} \{g \colon N \xrightarrow{\simeq} V, \ \hat{g} \colon \tau_N \cong g^* \tau_V\}$$

from the space of codimension zero PL immersions $f \colon N \to V$ that are homotopy equivalences to the space of PL homotopy equivalences $g \colon N \to V$ of $n$-manifolds, with a PL microbundle isomorphism $\hat{g} \colon \tau_N \cong g^* \tau_V$. It takes the PL immersion $f$ to the pair $(g, \hat{g}) = (f, \hat{f})$, where $f$ is viewed as a PL map, and $\hat{f} \colon \tau_N \cong f^* \tau_V$ is the induced PL microbundle isomorphism. By PL immersion theory [HP64, §2], this map is a weak homotopy equivalence. To apply immersion theory in codimension zero, as we are doing, we need to know that $N = M \times I \times J^k$ has no closed components, but this is clear. The usual formulation of immersion theory does not refer to $f$ and $g$ being homotopy equivalences, but these conditions just amount to restricting the immersion theory equivalence to some of the path components. Hence the fourth map is a weak homotopy equivalence.

The fifth map

$$\{g \colon N \xrightarrow{\simeq} V, \, \hat{g} \colon \tau_N \cong g^*\tau_V\} \xrightarrow{(5)} \{g \colon N \xrightarrow{\simeq} V, \, g^*\tau_V \text{ stably trivialized}\}$$

takes the PL microbundle isomorphism $\hat{g}$ to the associated stable isomorphism. When combined with the given stable framing of $N$, this specifies a stable trivialization of $g^*\tau_V$. For each $g$, the space of PL microbundle isomorphisms $\hat{g}$ can be identified with the space of sections in a principal $PL_n$-bundle over $N$, whereas the space of stable isomorphisms can be identified with the space of sections in the associated principal $PL$-bundle over $N$. By PL stability [HW65, Thm. 2] the stabilization map $PL_n \to PL$ is $(n-1)$-connected, and $N \simeq M$ has the homotopy type of a $d$-dimensional CW complex, with $n = d + 1 + k$. Hence the fifth map is at least $k$-connected, and this connectivity clearly grows to infinity with $k$.  $\square$

Recall from Definition 1.1.5 that a PL map of compact polyhedra is simple if all its point inverses are contractible, and from Proposition 2.1.3 that the composite of two simple PL maps is again simple. The simplicial categories in the following two definitions are the same as those introduced in Definition 3.1.4.

**Definition 4.1.8.** Let $s\mathcal{E}_\bullet$ be the simplicial category of compact PL bundles $\pi \colon E \to \Delta^q$ and their simple PL bundle maps, in simplicial degree $q$. Similarly, let $h\mathcal{E}_\bullet$ be the simplicial category with the same objects as $s\mathcal{E}_\bullet$, but with PL bundle maps over $\Delta^q$ that are homotopy equivalences as the morphisms, in simplicial degree $q$.

There are stabilization maps $\sigma \colon s\mathcal{E}_\bullet \to s\mathcal{E}_\bullet$ and $\sigma \colon h\mathcal{E}_\bullet \to h\mathcal{E}_\bullet$, given by product with the standard interval $J$, and a natural inclusion $j \colon s\mathcal{E}_\bullet \to h\mathcal{E}_\bullet$.

**Definition 4.1.9.** Let $s\widetilde{\mathcal{E}}_\bullet$ be the simplicial category of compact PL Serre fibrations $\pi \colon E \to \Delta^q$ and simple PL fiber maps over $\Delta^q$, in simplicial degree $q$. Similarly, let $h\widetilde{\mathcal{E}}_\bullet$ be the simplicial category with the same objects as $s\widetilde{\mathcal{E}}_\bullet$, but with PL homotopy equivalences over $\Delta^q$ as the morphisms, in simplicial degree $q$.

There are stabilization maps $\sigma \colon s\widetilde{\mathcal{E}}_\bullet \to s\widetilde{\mathcal{E}}_\bullet$ and $\sigma \colon h\widetilde{\mathcal{E}}_\bullet \to h\widetilde{\mathcal{E}}_\bullet$, given as before, and a natural inclusion $j \colon s\widetilde{\mathcal{E}}_\bullet \to h\widetilde{\mathcal{E}}_\bullet$. There are full embeddings $v \colon s\mathcal{E}_\bullet \subset s\widetilde{\mathcal{E}}_\bullet$ and $v \colon h\mathcal{E}_\bullet \subset h\widetilde{\mathcal{E}}_\bullet$ that view PL bundles as PL Serre fibrations. The functors $j$ and $v$ commute with one another, and with the stabilization maps $\sigma$.

The simplicial category $s\widetilde{\mathcal{E}}_\bullet^h(K)$ was introduced in Definition 1.1.6. There are forgetful maps

$$H^{PL}(M)^f \xrightarrow{\simeq} H^{PL}(M)^c \xrightarrow{u} s\widetilde{\mathcal{E}}_\bullet^h(M \times I),$$

where $u$ was introduced in Definition 1.1.7. We will also write $u$ for the composite map.

**Proposition 4.1.10.** *Let $K$ be a compact polyhedron. There is a homotopy fiber sequence*

$$s\widetilde{\mathcal{E}}^h_\bullet(K) \to s\widetilde{\mathcal{E}}_\bullet \xrightarrow{j} h\widetilde{\mathcal{E}}_\bullet$$

*based at $K$.*

This was proved as part of Proposition 3.3.1.

**Proposition 4.1.11.** *The vertical maps in the commutative square*

$$
\begin{array}{ccc}
s\mathcal{E}_\bullet & \xrightarrow{\ j\ } & h\mathcal{E}_\bullet \\
{\scriptstyle\simeq}\big\downarrow{\scriptstyle v} & & {\scriptstyle\simeq}\big\downarrow{\scriptstyle v} \\
s\widetilde{\mathcal{E}}_\bullet & \xrightarrow{\ j\ } & h\widetilde{\mathcal{E}}_\bullet
\end{array}
$$

*are homotopy equivalences. Hence the homotopy fiber of $j\colon s\mathcal{E}_\bullet \to h\mathcal{E}_\bullet$ at $K$ is homotopy equivalent to the homotopy fiber of $j\colon s\widetilde{\mathcal{E}}_\bullet \to h\widetilde{\mathcal{E}}_\bullet$ at $K$.*

This was proved as Corollary 3.5.3, as a consequence of Propositions 3.4.4 and 3.5.1.

**Lemma 4.1.12.** *The stabilization maps $\sigma\colon s\widetilde{\mathcal{E}}_\bullet \to s\widetilde{\mathcal{E}}_\bullet$, $\sigma\colon h\widetilde{\mathcal{E}}_\bullet \to h\widetilde{\mathcal{E}}_\bullet$ and*

$$\sigma\colon s\widetilde{\mathcal{E}}^h_\bullet(K) \to s\widetilde{\mathcal{E}}^h_\bullet(K \times J)$$

*are all homotopy equivalences. Hence so are the maps $\sigma\colon s\widetilde{\mathcal{E}}_\bullet \to \operatorname{colim}_n s\widetilde{\mathcal{E}}_\bullet$, $\sigma\colon h\widetilde{\mathcal{E}}_\bullet \to \operatorname{colim}_n h\widetilde{\mathcal{E}}_\bullet$ and*

$$\sigma\colon s\widetilde{\mathcal{E}}^h_\bullet(K) \to \operatorname*{colim}_k s\widetilde{\mathcal{E}}^h_\bullet(K \times I \times J^k).$$

*Proof.* The projection $pr\colon K \times J \to K$ is a simple PL map, hence gives a natural transformation from $\sigma\colon s\widetilde{\mathcal{E}}_\bullet \to s\widetilde{\mathcal{E}}_\bullet$ to the identity functor. Thus $\sigma$ is homotopic to the identity map. The case of $h\widetilde{\mathcal{E}}_\bullet$ is similar, and the case of $s\widetilde{\mathcal{E}}^h_\bullet(K)$ then follows from the homotopy fiber sequence of Proposition 4.1.10. The remaining conclusions follow by iteration. We permit stabilizing by $I$ once, in place of $J$, to pass from $K$ to $K \times I$. $\square$

We can now outline the proof of the manifold part of the stable parametrized $h$-cobordism theorem.

*Reduction of Theorem 1.1.8 to Theorem 4.1.14 and Proposition 4.1.15.* By the reduction made in Remark 1.1.9, we may assume that $M$ is a stably framed manifold. The forgetful map $u\colon \mathcal{M}^n_\bullet \to s\widetilde{\mathcal{E}}_\bullet$ factors as the composite

$$\mathcal{M}^n_\bullet \xrightarrow{\ w\ } s\mathcal{E}_\bullet \xrightarrow{\ v\ } s\widetilde{\mathcal{E}}_\bullet,$$

where $w$ forgets the stable framing and manifold structure, and $v$ views PL bundles as PL Serre fibrations. There is a similar factorization $u = vw$ with $h$-prefixes. Consider the following commutative diagram.

(4.1.13)

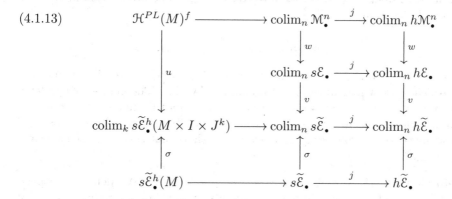

The vertical maps labeled $v$ and $\sigma$ are homotopy equivalences, by Proposition 4.1.11 and Lemma 4.1.12. The three complete rows are homotopy fiber sequences based at $M$, by Proposition 4.1.5 and Proposition 4.1.10. In each case the implicit null-homotopy of the composite map is given by the relevant structure map, i.e., a collar in an $h$-cobordism or a PL embedding and homotopy equivalence, so the homotopy fiber structures are compatible. To prove that the left hand vertical map $u$ is a homotopy equivalence, it therefore suffices to show that the middle and right hand vertical maps $w$ have this property. This is the content of the following two results. $\square$

**Theorem 4.1.14.** *The forgetful map*

$$w \colon \operatorname*{colim}_{n} \mathcal{M}_{\bullet}^{n} \to \operatorname*{colim}_{n} s\mathcal{E}_{\bullet}$$

*is a homotopy equivalence.*

This theorem is really the main geometric ingredient of the manifold part of the stable parametrized $h$-cobordism theorem, and will be proved in the next two sections. In contrast, the $h$-prefixed case is nearly trivial.

**Proposition 4.1.15.** *The forgetful map*

$$w \colon \operatorname*{colim}_{n} h\mathcal{M}_{\bullet}^{n} \to \operatorname*{colim}_{n} h\mathcal{E}_{\bullet}$$

*is a homotopy equivalence.*

*Proof.* We apply Theorem 3.6.7 (Theorem A$'$) and Addendum 3.6.10. The addendum applies because each PL bundle of polyhedra over a simplex is isomorphic to a product bundle. This time we emphasize left fibers. Let $K$ be any object of $h\mathcal{E}_0$, i.e., a polyhedron. Then $K$ embeds in some Euclidean $n$-space as a deformation retract of a regular neighborhood $N$, which is an object of $h\mathcal{M}_0^n$.

The transition maps of left fibers induced by $K \subset N \to K$ show that the left fiber $w/K$ is a retract of the left fiber $w/N$. The latter is a simplicial category with a terminal object in each degree, namely the identity map of the product bundle with fiber $N$, and is therefore contractible. The proposition follows, by the cited Theorem A'. $\square$

## 4.2. SPACES OF THICKENINGS

To prove Theorem 4.1.14, that $w \colon \operatorname{colim}_n \mathcal{M}^n_\bullet \to \operatorname{colim}_n s\mathcal{E}_\bullet$ is a homotopy equivalence, we will again employ the simplicial variant Theorem A' of Quillen's Theorem A. In this section we shall write $\operatorname{colim}_n w$ for this functor, for emphasis, and we will essentially recognize its homotopy fiber as a stabilized space of thickenings. In the following section we shall show that this stabilized space is contractible.

**Definition 4.2.1.** Let $K$ be a compact polyhedron. We define the space $S^n(K)$ of stably framed simple $n$-manifolds over $K$ as a simplicial set. By a **simple $n$-manifold over** $K$ we mean a compact PL $n$-manifold $M$ with a simple PL map $u \colon M \to K$. If $M$ is also equipped with a stable framing, then we call this a **stably framed** simple $n$-manifold over $K$. Let $S^n_0(K)$ denote the set of stably framed simple $n$-manifolds over $K$.

More generally, for a compact polyhedron $P$ we can consider a family of stably framed simple $n$-manifolds over $K$ parametrized by $P$, meaning a stably framed family of manifolds $\pi \colon E \to P$ parametrized by $P$, together with a simple PL map to $K \times P$ over $P$. In simplicial degree $q$ let $S^n_q(K)$ be the set of stably framed simple $n$-manifolds over $K$ parametrized by $\Delta^q$, i.e., the commutative diagrams

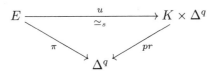

where $\pi \colon E \to \Delta^q$ is a stably framed family of manifolds parametrized by $\Delta^q$, and $u \colon E \to K \times \Delta^q$ is a simple PL map. For varying $q$, these sets assemble to the simplicial set $S^n(K)$.

There are stabilization maps $\sigma \colon S^n(K) \to S^{n+1}(K)$ given by taking $u \colon M \to K$ to $u \circ pr \colon M \times J \to K$, and similarly for parametrized families. Here $u \circ pr$ remains simple, because $pr \colon M \times J \to M$ is a simple PL map. As usual, $\sigma$ is a cofibration, so $\operatorname{colim}_n S^n(K)$ has the same homotopy type as the homotopy colimit.

The key fact that we need to establish is the following.

**Proposition 4.2.2.** *For each compact polyhedron $K$, the space $\operatorname{colim}_n S^n(K)$ is weakly contractible.*

We will translate this assertion to a statement about spaces of thickenings (Definition 4.2.4) later, in Lemma 4.2.5. That statement will follow immediately from Theorem 4.3.1. The final section of this book, Section 4.3, is devoted to the geometric proof of that theorem.

*Proof of Theorem 4.1.14, assuming Proposition 4.2.2.* It is not quite the case that $\mathrm{colim}_n S^n(K)$ equals the left fiber of $\mathrm{colim}_n w$ at $K$, so some maneuvering is required. We therefore introduce an auxiliary simplicial category $\mathcal{ME}_\bullet^n$ of (parametrized families of) $n$-manifolds mapping to polyhedra by simple PL maps. Let $\mathcal{ME}_0^n$ denote the category in which an object is a simple PL map

$$u \colon M \to K \, ,$$

where $M$ is a stably framed $n$-manifold and $K$ is a (now variable) compact polyhedron. A morphism from $u \colon M \to K$ to $u' \colon M \to K'$ is a commutative square

$$
\begin{array}{ccc}
M & \xrightarrow{\;u\;} & K \\
{\scriptstyle =}\big\downarrow & & \big\downarrow{\scriptstyle g} \\
M & \xrightarrow{\;u'\;} & K'
\end{array}
$$

where $g$ is a simple PL map. More generally, let $\mathcal{ME}_q^n$ be the category of simple PL bundle maps

$$u \colon E \to L$$

over $\Delta^q$, where $\pi \colon E \to \Delta^q$ is a stably framed family of $n$-manifolds as in Definition 4.1.2, and $\pi \colon L \to \Delta^q$ is a compact PL bundle as in Definition 4.1.8. A morphism from $u \colon E \to L$ to $u' \colon E \to L'$ is a commutative square

$$
\begin{array}{ccc}
E & \xrightarrow{\;u\;} & L \\
{\scriptstyle =}\big\downarrow & & \big\downarrow{\scriptstyle g} \\
E & \xrightarrow{\;u'\;} & L'
\end{array}
$$

of simple PL bundle maps over $\Delta^q$. In this way we obtain a simplicial category $\mathcal{ME}_\bullet^n$.

There are forgetful functors

$$\mathcal{M}_\bullet^n \xleftarrow{\;s\;} \mathcal{ME}_\bullet^n \xrightarrow{\;t\;} s\mathcal{E}_\bullet$$

that take $u \colon M \to K$ to the source $M$ and target $K$, respectively, and similarly in higher simplicial degrees. The source functor $s$ has a section $i \colon \mathcal{M}_\bullet^n \to \mathcal{ME}_\bullet^n$ given by mapping a stably framed $n$-manifold $M$ to the identity map $id \colon M \to M$, which is certainly a simple PL map to a polyhedron. But $s$ is also a deformation retraction, in view of the diagram

$$
\begin{array}{ccc}
M & \xrightarrow{\;=\;} & M \\
{\scriptstyle =}\big\downarrow & & \big\downarrow{\scriptstyle u} \\
M & \xrightarrow{\;u\;} & K \, ,
\end{array}
$$

which defines a natural transformation from $is$ to the identity functor of $\mathcal{ME}^n_\bullet$.

There is a stabilization map $\sigma\colon \mathcal{ME}^n_\bullet \to \mathcal{ME}^{n+1}_\bullet$ given by taking $u\colon M \to K$ to $u \circ pr\colon M \times J \to K$, just as for $S^n(K)$. The functor $s\colon \mathcal{ME}^n_\bullet \to \mathcal{M}^n_\bullet$ (but not its homotopy inverse $i$) is compatible with stabilization, so

$$\operatorname*{colim}_n s\colon \operatorname*{colim}_n \mathcal{ME}^n_\bullet \to \operatorname*{colim}_n \mathcal{M}^n_\bullet$$

is a homotopy equivalence, because at the level of homotopy groups,

$$\pi_*(\operatorname*{colim}_n s) = \operatorname*{colim}_n \pi_*(s)$$

is an isomorphism (for any choice of base point). It will therefore suffice for us to prove that the composite functor $\operatorname{colim}_n ws$ is a homotopy equivalence. The functor $w$ is compatible with the stabilization $\sigma$ on $s\mathcal{E}_\bullet$, while $t$ is only compatible with the identity stabilization on $s\mathcal{E}_\bullet$. To deal with this difference, and to avoid a homotopy coherence discussion, we will work at the level of homotopy groups.

There is a natural transformation

$$ws \to t\colon \mathcal{ME}^n_\bullet \to s\mathcal{E}_\bullet\,,$$

defined by the structural map of an object $u\colon M \to K$ of $\mathcal{ME}^n_\bullet$. Hence $ws$ and $t$ are homotopic, and $\pi_*(ws) = \pi_*(t)$. There is also a natural transformation $\sigma \to id\colon s\mathcal{E}_\bullet \to s\mathcal{E}_\bullet$, defined by the simple PL map $pr\colon K \times J \to K$. Hence $\sigma$ and $id$ are homotopic, and $\pi_*(\sigma) = id$. Thus the two squares induced at the level of homotopy groups by the two commuting squares

$$
\begin{array}{ccc}
\mathcal{ME}^n_\bullet & \xrightarrow{\ ws\ } & s\mathcal{E}_\bullet \\
\downarrow{\scriptstyle \sigma} & & \downarrow{\scriptstyle \sigma} \\
\mathcal{ME}^{n+1}_\bullet & \xrightarrow{\ ws\ } & s\mathcal{E}_\bullet
\end{array}
$$

and

$$
\begin{array}{ccc}
\mathcal{ME}^n_\bullet & \xrightarrow{\ t\ } & s\mathcal{E}_\bullet \\
\downarrow{\scriptstyle \sigma} & & \Vert \\
\mathcal{ME}^{n+1}_\bullet & \xrightarrow{\ t\ } & s\mathcal{E}_\bullet
\end{array}
$$

are equal. Hence the homomorphism

$$\operatorname*{colim}_n \pi_*(ws)\colon \operatorname*{colim}_n \pi_*(\mathcal{ME}^n_\bullet) \to \operatorname*{colim}_n \pi_*(s\mathcal{E}_\bullet)$$

(stabilization by $\sigma$ in the source and target) is equal to the homomorphism

$$\operatorname*{colim}_n \pi_*(t)\colon \operatorname*{colim}_n \pi_*(\mathcal{ME}^n_\bullet) \to \operatorname*{colim}_n \pi_*(s\mathcal{E}_\bullet) = \pi_*(s\mathcal{E}_\bullet)$$

(stabilization by $\sigma$ in the source, by $id$ in the target). Thus the functor $\operatorname{colim}_n w$ of the theorem is a homotopy equivalence if and only if the functor

$$\operatorname{colim}_n t \colon \operatorname{colim}_n \mathcal{ME}_\bullet^n \to s\mathcal{E}_\bullet$$

is a homotopy equivalence.

We apply Theorem 3.6.7 (Theorem A') to $\operatorname{colim}_n t$. Each object of $s\mathcal{E}_\bullet$ is 0-dimensional up to isomorphism, so we may apply Addendum 3.6.10 to deduce that $\operatorname{colim}_n t$ is a homotopy equivalence if its left fiber at each object $K$ of $s\mathcal{E}_0$ is contractible. Unraveling the definitions, the left fiber $t/K$ of $t \colon \mathcal{ME}_\bullet^n \to s\mathcal{E}_\bullet$ at $K$ is easily seen to deformation retract onto the space $S^n(K)$ of stably framed simple $n$-manifolds over $K$, in such a way that the stabilization maps agree. Hence the left fiber of $\operatorname{colim}_n t$ at $K$ is homotopy equivalent to $\operatorname{colim}_n S^n(K)$. Therefore Proposition 4.2.2 implies Theorem 4.1.14. $\quad\square$

*Remark 4.2.3.* At first glance, one might hope to prove Proposition 4.2.2 by showing that the connectivity of the spaces $S^n(K)$ grows to infinity with $n$. However, this is definitely not true. In the case $K = *$, a simple $n$-manifold $M$ over $*$ is just a contractible (compact PL) manifold. Poincaré duality implies that $M$ is a homotopy ball in the weak sense, i.e., the boundary $\partial M$ has the homology of an $(n-1)$-sphere, but it is not necessarily simply-connected. Indeed, for $n \geq 5$ many (perfect) fundamental groups $\pi_1(\partial M)$ arise in this way, so $S^n(*)$ is not even path-connected. For this reason, it is apparently not possible to prove the stability theorem for PL concordances (or $h$-cobordisms) by this method. Still, this pathology is an unstable one, and can be avoided in the stable context by the refined definition below.

**Definition 4.2.4.** Let $K$ be a compact polyhedron. We define the space $T^n(K)$ of stably framed $n$-manifold thickenings of $K$ as a simplicial subset of $S^n(K)$. An $n$-**manifold thickening** of $K$ shall consist of a compact PL $n$-manifold $M$ and a PL map $u \colon M \to K$ that is an **unthickening map**, in the sense that:

(1) $u \colon M \to K$ has contractible point inverses, i.e., is a simple PL map; and

(2) the restricted map $u|\partial M \colon \partial M \to K$ has 1-connected point inverses.

(By definition, a space is 1-connected if and only if it is non-empty, path-connected and simply-connected.)

By insisting that the manifold is stably framed, we obtain the notion of a **stably framed** $n$-**manifold thickening**. These are the 0-simplices of $T^n(K)$. By working with parametrized families over affine simplices, we obtain the simplicial set $T^n(K)$. The stabilization map $\sigma \colon T^n(K) \to T^{n+1}(K)$ takes $u \colon M \to K$ to the thickening $u \circ pr \colon M \times J \to K$.

The word "thickening" has also been used with slightly different definitions elsewhere in the literature. Our usage is similar to that of [Wa66, §1], but a little more restrictive.

**Lemma 4.2.5.** *The inclusion*

$$\operatorname*{colim}_{n} T^{n}(K) \to \operatorname*{colim}_{n} S^{n}(K)$$

*is an isomorphism, where the maps in the direct system are given by the stabilization maps $\sigma$.*

Together with Proposition 4.2.8, which follows shortly, this lemma makes precise how the problem discussed in Remark 4.2.3 goes away after stabilization. It will be proved after the following two lemmas.

**Lemma 4.2.6.** *A map $g\colon X \to Y$ of simplicial sets, whose geometric realization has 1-connected point inverses $|g|^{-1}(p)$ for all $p \in |Y|$, induces a bijection $\pi_0(g)\colon \pi_0(X) \to \pi_0(Y)$ and an isomorphism $\pi_1(g)\colon \pi_1(X) \to \pi_1(Y)$, for each choice of base point in $X$.*

*Proof.* This amounts to an elementary check in terms of the usual edge-path presentation of the fundamental group. In fact it suffices that $|g|^{-1}(p)$ is 1-connected for each 0-simplex $p$ in $Y$, that $|g|^{-1}(p)$ is path-connected for some $p$ in the interior of each 1-simplex of $Y$, and that $|g|^{-1}(p)$ is non-empty for some $p$ in the interior of each 2-simplex of $Y$. To prove, for example, that $\pi_1(g)$ is injective, consider a closed loop $\gamma$ of 1-simplices in $X$ representing an element in the kernel of $\pi_1(g)$, choose a null-homotopy of its image under $g$, and deform it to a disc of 2-simplices in $Y$. Choose lifts in $X$ of each of the 2-simplices occurring in this disc, choose rectangles in $X$ connecting the pairs of edges that were identified in $Y$, and choose null-homotopies in $X$ of the loops of rectangle edges that lie over vertices in $Y$. These choices glue together to give a null-homotopy of the given closed loop $\gamma$. The proofs of the remaining claims are easier, and will be omitted. $\square$

**Lemma 4.2.7.** *The composite $fu$ of an unthickening map $u\colon M \to K$ with a simple PL map $f\colon K \to L$ is again an unthickening map.*

*Proof.* The composite $fu$ is a composite of simple PL maps, and is therefore simple. For each point $p \in L$ the restricted map $u'\colon (fu|\partial M)^{-1}(p) \to f^{-1}(p)$ has 1-connected point inverses, because $u$ is an unthickening map, and the target $f^{-1}(p)$ is contractible, because $f$ is simple. Triangulating $u'$ by a map $g\colon X \to Y$ of simplicial sets, we deduce from Lemma 4.2.6 that $(fu|\partial M)^{-1}(p)$ is 1-connected, so $fu$ is indeed an unthickening map. $\square$

*Proof of Lemma 4.2.5.* The projection $pr\colon M \times J^3 \to M$ is an unthickening map, so composition with $pr$ takes any simple PL map $u\colon M \to K$ in $S^n(K)$ to an unthickening map $u \circ pr\colon M \times J^3 \to K$ in $T^{n+3}(K)$, by Lemma 4.2.7. After stabilization, the resulting maps $S^n(K) \to T^{n+3}(K)$ and $T^n(K) \subset S^n(K)$ become mutual inverses. $\square$

In view of the lemma just proved, the following proposition implies the special case $K = *$ of Proposition 4.2.2, and will be a basic building block in its proof.

**Proposition 4.2.8.** *For $n \geq 6$ the space $T^n(*)$ is the space of stably framed PL $n$-balls, which is $(n-2)$-connected.*

*Proof.* An $n$-manifold thickening $u \colon M \to *$ (of $K = *$) is the same as a contractible $n$-manifold with 1-connected boundary, i.e., a homotopy $n$-ball in the strong sense. Provided that $n \geq 6$ it is therefore an honest PL $n$-ball, in view of the $h$-cobordism theorem of Smale. Therefore the space $T^n(*)$ of stably framed $n$-manifold thickenings is the same as the space of PL $n$-balls with a stable microbundle trivialization, which is a homotopy fiber of the map

$$BPL(D^n, \partial D^n) \to BPL = \operatorname*{colim}_n BPL_n .$$

Here $PL(D^n, \partial D^n)$ is the simplicial group of PL automorphisms of $D^n$ that are not required to fix the boundary, $PL_n$ is the simplicial group of germs (near 0) of PL automorphisms of $\mathbb{R}^n$, and $PL = \operatorname{colim}_n PL_n$ over the standard stabilization maps $\sigma \colon PL_n \to PL_{n+1}$. This space of stably framed PL $n$-balls is $(n-2)$-connected. We give some details, following Haefliger–Wall [HW65, §4].

The map $PL(D^n, \partial D^n) \to PL(S^{n-1})$ that restricts a PL automorphism to the boundary $\partial D^n = S^{n-1}$ is a homotopy equivalence, by the Alexander trick. Furthermore, there is a homotopy commutative diagram

$$
\begin{array}{ccccc}
PL(S^{n-1} \operatorname{rel} s_0) & \longrightarrow & PL(S^{n-1}) & \longrightarrow & S^{n-1} \\
\Big\downarrow{\simeq} & & \Big\downarrow & & \\
PL_{n-1} & \xrightarrow{\ \sigma\ } & PL_n & \xrightarrow{\ \sigma\ } & PL
\end{array}
$$

where $s_0 \in S^{n-1}$ is a base point, and the upper row is the fiber sequence that results from evaluating each PL automorphism of $S^{n-1}$ at $s_0$. The left hand vertical map is a homotopy equivalence, by the PL version [KL66, Thm. 1] of the "microbundles are fiber bundles" theorem of J. M. Kister and B. Mazur. The lower horizontal maps are $(n-2)$-connected and $(n-1)$-connected, respectively, by the PL stability theorem of Haefliger–Wall [HW65, Thm. 2]. Thus the composite map $PL(S^{n-1}) \to PL_n \to PL$ is $(n-2)$-connected, which just means that the space $T^n(*)$ is $(n-2)$-connected. $\square$

### 4.3. STRAIGHTENING THE THICKENINGS

Saving the best for last, we show in this section that the connectivity of thickening spaces grows to infinity with the manifold dimension, so that the stabilized thickening spaces are weakly contractible. In view of Lemma 4.2.5 this concludes the proof of Proposition 4.2.2, which was the final reduction of Theorem 1.1.8.

Let $K$ be a compact polyhedron, and recall that $T^n(K)$ denotes the space of stably framed $n$-manifold thickenings of $K$.

**Theorem 4.3.1.** *The space $T^n(K)$ is at least $(n - 2k - 6)$-connected, where $k$ is the dimension of $K$ and $n \geq 2k + 6$.*

The proof proceeds, essentially, by an induction on the dimension of the polyhedron $K$. As is customary with such an inductive procedure, we have to generalize it by putting it into a relativized form, so that, during the construction, we will be able to keep track of things done previously. The generalization consists of working relative to a subpolyhedron. For technical reasons, it is convenient to insist that the polyhedron itself has a special structure relative to the subpolyhedron, namely that it is a PL mapping cylinder.

*Remark 4.3.2.* We review the various mapping cylinders in the polyhedral context. Let $f \colon L \to K$ be a PL map of compact polyhedra. The construction of the usual mapping cylinder $L \times \Delta^1 \cup_L K$ does not give a polyhedron, unless $f$ is (locally) injective, so one must modify the construction. Following J. H. C. Whitehead [Wh39], this can be done by first triangulating $f$ by a simplicial map $g \colon X \to Y$, and then building a **PL mapping cylinder** $W_f$ by starting with the disjoint union $L \coprod K$ and inductively attaching a cone on $s \cup W_{f|\partial s} \cup f(s)$ for each affine simplex $s$ of $L$, in some order of increasing simplex dimension.

Here $f(s)$ denotes the image of $s$ under $f$, viewed as an affine simplex of $K$, and $f|\partial s$ denotes the restricted map $\partial s \to f(\partial s)$, from the boundary of $s$ to its image under $f$. These all inherit triangulations from $g \colon X \to Y$. If $s$ is the geometric realization of a non-degenerate simplex $x$ in $X$, then $f(s)$ is the geometric realization of the non-degenerate part $g^{\#}(x)$ of $g(x)$. If $Z \subset X$ is the simplicial subset generated by the proper faces of $x$, then $f|\partial s$ is triangulated by $g|Z \colon Z \to g(Z)$.

Near the front end, Whitehead's PL mapping cylinder $W_f$ contains a copy of the product $L \times \Delta^1$, triangulated in such a way that it contains $X$ and $Sd(X)$ at the two ends. M. Cohen [Co67, §4] gives a construction of the remaining piece of $W_f$, denoted $C_f$, by starting with $K$ and then inductively attaching a cone on $\partial s \cup C_{f|\partial s} \cup f(s)$ for each affine simplex $s$ of $L$, in some order of increasing simplex dimension. At the front end, Cohen's PL mapping cylinder $C_f$ contains a copy of $L$ triangulated as $Sd(X)$, and there is a canonical PL homeomorphism $W_f \cong L \times \Delta^1 \cup_L C_f$.

There is also a non-canonical PL homeomorphism $W_f \cong C_f$ [Co67, Cor. 9.4]. By [Wh39, §10], there is a TOP homeomorphism $W_f \cong L \times \Delta^1 \cup_L K$ relative to $L \coprod K$, so there is also a TOP homeomorphism $C_f \cong L \times \Delta^1 \cup_L K$ relative to $L \coprod K$, but we shall not rely on these results.

As described, the PL mapping cylinders $W_f$ and $C_f$ depend on the chosen triangulation $g \colon X \to Y$ of $f \colon L \to K$. By [Co67, Prop. 9.5] the PL structure on $W_f$ and $C_f$ is abstractly invariant under subdivision of $g$, but only up to non-canonical isomorphism. Thus the PL isomorphism class of the PL mapping cylinders $W_f$ and $C_f$ is well defined for a given PL map $f \colon L \to K$, but the rule $f \mapsto C_f$, of choosing a representative in this isomorphism class, can not be made functorial.

There is a subdivision of $C_f$ obtained by starring at the barycenter of each non-degenerate simplex in $K \subset C_f$, in order of decreasing dimension. From its description in [Co67, p. 225], see also [Ak72, p. 408], it equals the nerve of the partially ordered set of non-degenerate simplices of $X$ and $Y$, partially ordered by inclusion within $X$ and $Y$, and by the relation $x > g^{\#}(x)$ between simplices in $X$ and $Y$. This is precisely our reduced mapping cylinder, in the backward version, of the normally subdivided map $Sd(g) \colon Sd(X) \to Sd(Y)$. In symbols:

$$C_{|g|} = |M(Sd(g))| \,.$$

In particular, this construction is functorial in $g$. We shall also make use of the explicit product structure near $L$ in $C_f$, so in effect we are working with Whitehead's PL mapping cylinder $W_f$.

**Definition 4.3.3.** Let $f \colon L \to K$ be a PL map of polyhedra. Choose a triangulation $g \colon X \to Y$ of $f$, and let

$$W_f = L \times [0,1] \cup_{L \times 1} C_f$$

be Whitehead's PL mapping cylinder for $f = |g|$, with the explicit collar $L \times [0,1]$ near the front end $L \cong L \times 0 \subset W_f$. We shall sometimes denote $W_f$ by $W(L \to K)$.

Returning to Theorem 4.3.1 and its relativization, the base polyhedron $K$ will henceforth be replaced by a choice of PL mapping cylinder $W_f$ for a PL map $f \colon L \to K$. It contains a collared copy of $L$ at the front end, and a copy of $K$ at the back end. We shall work relative to a fixed stably framed $(n-1)$-manifold thickening

$$v \colon N \to L \,,$$

and we shall consider the space of stably framed $n$-manifold thickenings

$$u \colon M \to W_f \,,$$

where $M$ contains $N$ in its boundary $\partial M$ and $u$ restricted to $N$ is the fixed stably framed thickening $v$. See Figure 2, where the arrows labeled $M$ refer the whole of the upper part of the figure, and the arrows labeled $W_f$ refer to the whole of the lower part. For technical reasons, we shall also want the thickening $u$ to be given as a product with the thickening $v$ in a neighborhood of $N$ in $M$, as we now specify.

**Definition 4.3.4.** We define the space $T^n(L \to K, N)$ of stably framed $n$-manifold **thickenings over $L \to K$ relative to $N$** as a simplicial set. A 0-simplex consists of:

(1) a stably framed $n$-manifold thickening $u \colon M \to W_f$ of the chosen PL mapping cylinder $W_f$; and

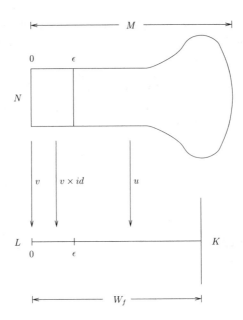

FIGURE 2. An unthickening map $u\colon M \to W_f$, relative to the fixed unthickening $v\colon N \to L$

(2) a stably framed PL embedding $c\colon N \times [0, \epsilon] \to M$, with $0 < \epsilon \le 1$, such that the diagram

$$
\begin{array}{ccccc}
N & \rightarrowtail & N \times [0, \epsilon] & \overset{c}{\rightarrowtail} & M \\
\downarrow{\scriptstyle v} & & \downarrow{\scriptstyle v \times id} & & \downarrow{\scriptstyle u} \\
L & \rightarrowtail & L \times [0, \epsilon] & \rightarrowtail & W_f
\end{array}
$$

commutes, and the image of $N \times [0, \epsilon)$ in $M$ equals the preimage under $u$ of $L \times [0, \epsilon)$ in $W_f$, hence is open in $M$. Here $v$ is the fixed thickening of $L$, the remaining horizontal maps are the obvious embeddings, and $N \times [0, \epsilon]$ is given the product stable framing from $N$ and $[0, \epsilon] \subset \mathbb{R}$.

A $q$-simplex in $T^n(L \to K, N)$ consists of:

(1) a stably framed family

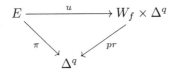

of $n$-manifold thickenings of $W_f$ parametrized by $\Delta^q$; and

(2) a PL family of stably framed embeddings

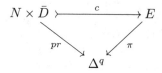

where $\bar{D} \subset [0,1] \times \Delta^q$ consists of the $(t,p)$ with $0 \leq t \leq \epsilon(p)$, $p \in \Delta^q$, for some PL function $\epsilon \colon \Delta^q \to (0,1]$, making the analog of the rectangular diagram above commute, and such that $c(N \times D) = u^{-1}(L \times D)$, where $D \subset \bar{D}$ consists of the $(t,p)$ with $0 \leq t < \epsilon(p)$.

In other words, the thickening of $W_f$ is locally constant in the collar coordinate, near the preimage $N$ of $L$, but the extent of the locally constant region (as given by $\epsilon$) may vary within a parametrized family.

It is a consequence of these hypotheses that $N \times 0 \cup \partial N \times [0,\epsilon)$ embeds as an open subset of $\partial M$, because the image of $N \times [0,\epsilon)$ is open in $M$.

The notation is slightly imprecise, because $T^n(L \to K, N)$ really depends on the choice of PL mapping cylinder $W_f$ and on the unthickening map $v \colon N \to L$.

All thickenings to be considered from here on will be stably framed, even if we sometimes omit to mention it.

Replacing $\Delta^q$ by a compact polyhedron $P$ we obtain the notion of a family of thickenings over $L \to K$, relative to $N$ and parametrized by $P$. We will write $u \colon M_p = \pi^{-1}(p) \to W_f$ for an individual thickening of $W_f$ that occurs as part of such a family, at a variable point $p \in P$.

Here is the relative form of Theorem 4.3.1, to be proved in this section. It includes Theorem 4.3.1 as the special case $L = \emptyset$, when $K = W_f$.

**Theorem 4.3.5.** *The space $T^n(L \to K, N)$ is at least $(n - 2w - 6)$-connected, where $w = \dim(W_f) = \max(\dim(L) + 1, \dim(K))$ and $n \geq 2w + 6$.*

The following amounts to the same thing.

**Reformulation of Theorem 4.3.5.** *Let $P$ be a compact polyhedron of dimension at most $(n - 2w - 6)$. Then any PL family of stably framed $n$-manifold thickenings over $L \to K$, relative to $N$ and parametrized by $P$, may be deformed into a constant family.*

By a **deformation** of such a family over $P$ we mean a family over $P \times [0,1]$ that restricts to the given family over $P \cong P \times 0$. It is a deformation **into a constant family** when the resulting family over $P \cong P \times 1$ is constant in its dependence on $p \in P$. We shall also call a deformation into a constant family a **straightening** of the family, hence the title of this section.

The theorem implies that under the given dimensional hypotheses, any one family over $P$ may be deformed into any other family, constant or not. For by viewing the two families as a single family over $P \coprod P$, which has the same dimension as $P$, and straightening the combined family, we get deformations connecting both the first and the second family to the same constant family. The obvious composite deformation then connects the first family to the second.

*Proof of Theorems 4.3.1 and 4.3.5.* The proof proceeds by induction on the dimension $k = \dim(K)$ of the target polyhedron $K$, **not** the dimension of the PL mapping cylinder $W_f$ or of the source polyhedron $L$. Both the beginning of the induction $(k = 0)$ and the inductive steps $(k > 0)$ are nontrivial, and they require rather different arguments. With either of them we shall start by treating the special (absolute) case when $L = \emptyset$, and then proceed to the general (relative) case when $L \neq \emptyset$.

There is only something to prove if $P$ is non-empty and $n \geq 2w + 6$, so we will assume this. In particular $n \geq 6$, and $n \geq 8$ in the relative cases. The precise dimension estimate could perhaps be improved with a more complicated formula, but basically we need $n \geq 6$ to apply the $s$-cobordism theorem, and we lose twice the dimension of $L$ in a comparison between embeddings and immersions, so the estimate $n \geq 2w + 6$ is about the simplest one that works.

### 4.3.6. Inductive beginning, the absolute case.

This part is the case $\dim(K) = 0$ and $L = \emptyset$, and uses the Haefliger–Wall PL stability theorem [HW65].

Without loss of generality we may assume that $K$ is connected, so $K = *$ is a point. Then $T^n(\emptyset \to *, \emptyset) = T^n(*)$, and this space is $(n-2)$-connected for $n \geq 6$, as we already deduced in Proposition 4.2.8 of the previous section.

### 4.3.7. Inductive beginning, the relative case.

This part is the case $\dim(K) = 0$ and $L$ non-empty, and uses general position and the Haefliger–Poenaru PL immersion theory [HP64]. It also uses the absolute case just considered.

We may again assume that $K$ is connected, so $K = *$. The PL mapping cylinder $W_f = W(L \to *)$ is a cone in this case. More precisely, $C_f = \mathrm{cone}(L)$ with vertex $*$, and

$$W_f = L \times [0,1] \cup_{L \times 1} \mathrm{cone}(L).$$

Then there is a PL 1-parameter family of self-maps

$$g_t \colon W_f \to W_f$$

for $t \in [0,1]$, with the following properties, where $\epsilon$ is some fixed number in $(0,1]$.

(1) $g_t$ is a simple PL map and its restriction to $L \times [0, \epsilon/2]$ is the identity, for each $t \in [0,1]$.

(2) $g_0$ is the identity map, and $g_1$ maps $L \times [\epsilon, 1] \cup_{L \times 1} \mathrm{cone}(L)$ to the cone vertex $*$ and $L \times [0, \epsilon)$ to the complement of $*$.

In other words, there is a PL map $G \colon W_f \times [0,1] \to W_f$ such that $g_t(x) = G(x,t)$ has these properties. To prove this, it suffices to give formulas for such a family of self-maps in the case where $L$ is a simplex, in a way that is natural with respect to inclusion maps of ordered simplices. We omit the formulas.

Recall from Definition 4.3.4 that over a neighborhood of $L$, any parametrized family of thickenings is required to be the constant extension in the collar coordinate of the standard thickening $v \colon N \to L$. Let $\epsilon$ be so small that $L \times [0, \epsilon)$ is such a neighborhood for the particular family parametrized by $P$ that we are considering. Composing the structure map $u$ of the thickenings in that family with the self-maps $g_t$ above, we obtain a deformation $\{g_t u\}_t$ of the family. It is a deformation through thickenings by Lemma 4.2.7 above, because each $g_t$ is simple. The result of the deformation, $g_1 u$, is a new family of thickenings parametrized by $P$, which admits the following description. Letting $p \in P$ denote a variable point, we may write

$$M_p = N \times [0, \epsilon] \cup_N M_p',$$

where the attaching at the $\epsilon$-end is by means of a stably framed PL embedding $N \to \partial M_p'$, and the structural map $u \colon M_p \to W_f$ is essentially independent of $p$. More precisely, on the part $N \times [0, \epsilon]$ the structural map is some fixed map that only depends on the original map $v \colon N \to L$ and the homotopy $g_t$ (namely $g_1 \circ (v \times id)$), and all of $M_p'$ is mapped to $*$.

With the family $p \mapsto M_p'$ we are in the case treated before: It is a family of (stably framed $n$-manifold) thickenings of $K = *$. When considered relative to nothing, it may thus be deformed to a constant family, because $\dim(P) \leq (n-2)$ and $n \geq 6$. In other words, $p \mapsto M_p'$ is a trivializable PL $n$-ball bundle. The attaching data are now equivalent to a map from $P$ to the space of stably framed embeddings

$$N \to S^{n-1}.$$

We have to check that this space is at least $(n - 2w - 6)$-connected. Since $P$ is assumed to be non-empty, at least one such stably framed codimension zero embedding of $N$ in $S^{n-1}$ exists.

Let $h$ denote the **homotopical dimension** of $N$, i.e., the minimal integer $h$ such that there exists a handle decomposition of $N$ with all handles of index $\leq h$. The homotopical dimension does not increase under stabilization, i.e., upon replacing $N$ by $N \times J$. (It can decrease, as illustrated by [Po60] and [Ma61].)

**Lemma 4.3.8.** *The homotopical dimension $h$ of $N$ is at most equal to $\dim(L)$.*

*Proof.* There is a map $e \colon L \to N$ that is homotopy inverse to the unthickening map $v \colon N \to L$, because a simple map is a homotopy equivalence. By general position, this map $e$ may be assumed to be a PL embedding into the interior of $N$, because $2 \dim(L) < \dim(N)$ by the dimensional hypothesis $n \geq 2w + 6$. Let $R$ be a regular neighborhood (contained in the interior of $N$) of the embedded image of $L$ in $N$, and let $W$ be the part of $N$ not in the interior of $R$. Then $N = R \cup_{\partial R} W$ is a union of two codimension zero submanifolds, and $\partial W =$

$\partial R \coprod \partial N.$

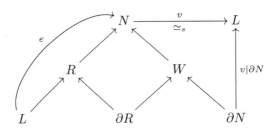

The regular neighborhood $R$ has homotopical dimension at most $\dim(L)$, because a handle decomposition for it may be constructed out of a triangulation of $L$. See, e.g., [RS72, Prop. 6.9]. We claim that $W$ is an $s$-cobordism from $\partial R$ to $\partial N$. The dimensional hypothesis ensures that $\dim(\partial N) \geq 6$, so the classical $s$-cobordism theorem will then imply that $W$ is a product. This implies that no new handles are needed to build $N$ from $R$, so the homotopical dimension of $N$ will be the same as that of $R$.

It remains to prove the claim. The regular neighborhood inclusion $L \subset R$ is an expansion. In particular it is a simple homotopy equivalence, i.e., a homotopy equivalence with zero Whitehead torsion. The homotopy inverse $e$ to the simple map $v$ is also a simple homotopy equivalence, hence so is the factor $R \subset N$, by the composition law for Whitehead torsion [Co73, 20.2].

By uniqueness of regular neighborhoods, and [Co67, Cor. 9.7], $R$ is homeomorphic to the PL mapping cylinder of a map $\partial R \to L$. Hence $R \backslash L$ deformation retracts to $\partial R$. We have $\dim(L) \leq \dim(R) - 3$ by the dimensional hypothesis again, so $\partial R \subset R$ induces bijections on $\pi_0$ and $\pi_1$ by general position. In the same way, $N \setminus L$ deformation retracts to $W$, and $W \subset N$ induces bijections on $\pi_0$ and $\pi_1$. The assumption that $v$ is an unthickening map implies that $v|\partial N \colon \partial N \to L$ induces bijections on $\pi_0$ and $\pi_1$, by Lemma 4.2.6. It follows that $\partial N \subset W$ also induces bijections on $\pi_0$ and $\pi_1$, which is part of the requirement for $W$ to be an $s$-cobordism.

The inclusion $R \subset N$ is homotopy equivalence, hence a homology equivalence with arbitrary local coefficients. By excision, and the $\pi_0$- and $\pi_1$-bijections $\partial R \subset R$ and $W \subset N$ just established, $\partial R \subset W$ is also a homology equivalence with arbitrary local coefficients, hence a homotopy equivalence. The same $\pi_0$- and $\pi_1$-bijections, and the excision lemma for Whitehead torsion [Co73, 20.3], imply that the Whitehead torsion of $\partial R \subset W$ equals the Whitehead torsion of $R \subset N$, i.e., is zero. Hence $\partial R \subset W$ is a simple homotopy equivalence. This is the remaining requirement for $W$ to be an $s$-cobordism from $\partial R$ to $\partial N$. $\square$

**Lemma 4.3.9.** *The space of stably framed embeddings $N \to S^{n-1}$ is at least $(n - 2h - 3)$-connected.*

*Proof.* By general position for handle cores, the inclusion

$$\{stably\ framed\ embeddings\ N \to \mathbb{R}^{n-1}\}$$

$$\to \{stably\ framed\ embeddings\ N \to S^{n-1}\}$$

induced by composition with the one-point compactification $\mathbb{R}^{n-1} \to S^{n-1}$ is $(n - h - 2)$-connected. In particular it is a $\pi_0$-bijection, so that $N$ admits a stably framed codimension zero embedding in $\mathbb{R}^{n-1}$. By selecting one of these, we obtain a framing $\tau_N \cong \epsilon^{n-1}$ of $N$ that is compatible with the given stable framing of $N$. Furthermore, the existence of such an embedding implies that $\partial N$ meets each path component of $N$, as required for the use of codimension zero immersion theory that is soon to follow.

By general position again, the forgetful map

$$\{stably\ framed\ embeddings\ N \to \mathbb{R}^{n-1}\}$$

$$\to \{stably\ framed\ immersions\ N \to \mathbb{R}^{n-1}\}$$

is $(n - 2h - 2)$-connected, because an immersion of $N$ without double points is an embedding. By immersion theory, [HP64], the "differential"

$$\{immersions\ N \to \mathbb{R}^{n-1}\} \xrightarrow{d} \{maps\ N \to PL_{n-1}\}$$

is a homotopy equivalence. Here we have used the selected framing of $N$ to identify the PL microbundle maps $\tau_N \to \epsilon^{n-1}$ with maps $N \to PL_{n-1}$. Consequently, the map

$$\{stably\ framed\ immersions\ N \to \mathbb{R}^{n-1}\} \to \{maps\ N \to \mathrm{hofib}(PL_{n-1} \to PL)\}$$

(of homotopy fibers over the space of maps $N \to PL$, which compares the stable framings of $N$ and $\mathbb{R}^{n-1}$) is also a homotopy equivalence.

By the stability theorem for $PL_{n-1} \to PL$, [HW65], the target space in the previous display is $(n - h - 3)$-connected. Hence the space

$$\{stably\ framed\ embeddings\ N \to S^{n-1}\}$$

is at least $(n - 2h - 3)$-connected.   □

Now $2h + 3 \le 2\dim(L) + 3 < 2w + 6$, so any map from $P$ to the space of stably framed embeddings $N \to S^{n-1}$ can, indeed, be deformed to a constant map.

### 4.3.10. Inductive step, the absolute case.

This part is the case $\dim(K) > 0$ and $L = \emptyset$, so $K = W_f$. We assume, by induction, that the reformulation of Theorem 4.3.5 holds for thickenings over $L^* \to K^*$ where $\dim(K^*) < \dim(K)$.

The argument relies upon a principle of **global transversality in patches**, which was invented by Hatcher for the purpose of proving a theorem on 3-manifolds [Ha76, Lem. 1]. See also [Ha99, Thm. 1]. In the latter preprint Hatcher works differentiably (as is the customary way of discussing transversality arguments). We will instead have to discuss transversality in the PL context (sorry, we have to), as Hatcher did in his aforementioned paper.

When the base polyhedron $K$ has dimension $k = \dim(K) > 0$, it may be decomposed as

$$K = K_1 \cup_{K_0} K_2$$

where each of $K_1$ and $K_2$ can be regarded as the PL mapping cylinder of some PL map with source $K_0$ and with target of dimension smaller than the dimension of $K$. The pieces are thus regarded as simpler, in our inductive scheme, than $K$ itself. An example of such a decomposition can be obtained from any triangulation of $K$, by letting $K_1$ consist of the top-dimensional simplices, trimmed down to half size, and taking $K_2$ to be the closure of the complement of $K_1$, which is a regular neighborhood of the $(k-1)$-skeleton of $K$. These meet along the boundary $K_0$ of $K_1$, which is the disjoint union of one PL $(k-1)$-sphere for each $k$-simplex of $K$.

With an eye to the transversality argument to be given shortly, we note that $K_0$ is actually bicollared in $K$. We emphasize this fact by rewriting the decomposition as

$$K = K_1 \cup_{K_0} (K_0 \times [0,1]) \cup_{K_0} K_2 \,.$$

In more detail, if $K'$ is the discrete set of barycenters of the non-degenerate $k$-simplices of $K$, and $K''$ is the $(k-1)$-skeleton of $K$, then $K_1$ and $K_2$ may be regarded as the PL mapping cylinders $W(K_0 \to K')$ and $W(K_0 \to K'')$, respectively. More generally, such a decomposition can be obtained from any cell decomposition of $K$.

The fact underlying Hatcher's principle is the following lemma. Since parametrized PL transversality is a somewhat delicate notion, we give the proof in some detail. Let

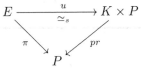

be a PL family of stably framed $n$-manifold thickenings of $K$, parametrized by $P$, and let $M_p = \pi^{-1}(p)$ for each $p \in P$.

**Lemma 4.3.11.** *There are a finite number of subpolyhedra $P_i \subset P$, whose interiors cover $P$, and real numbers $t_i \in (0,1)$, such that the thickenings $u \colon M_p \to K$ in the subfamily parametrized by $P_i$ are "PL transverse" to the bicollared subpolyhedron $K_0 \times t_i$ of $K$, in the sense that:*

*(a) the fiber products*

$$p \mapsto N_p := M_p \times_K (K_0 \times t_i)$$

*for $p \in P_i$ form a PL family of stably framed $(n-1)$-manifold thickenings of $K_0 \times t_i \cong K_0$, and*

*(b) for $\epsilon > 0$ sufficiently small there is a PL isomorphism from the product family*

$$p \mapsto N_p \times [t_i - \epsilon, t_i + \epsilon]$$

*over $P_i$ to the fiber product family*

$$p \mapsto M_p \times_K \left( K_0 \times [t_i - \epsilon, t_i + \epsilon] \right),$$

*also over $P_i$, which commutes with the projection to $[t_i - \epsilon, t_i + \epsilon]$.*

*Proof.* By assumption, $\pi \colon E \to P$ is a PL bundle, so for each point $p \in P$ we can find a neighborhood $U \subset P$ of $p$ and a PL isomorphism $M_p \times U \cong E_U = \pi^{-1}(U)$ over $U$. As usual we may take $U$ to be a compact subpolyhedron of $P$, and to simplify the notation we assume that $U = P$. Let $b \colon K \to I = [0,1]$ map $K_1$ to 0, $K_2$ to 1 and project $K_0 \times [0,1]$ to the second (bicollar) coordinate. Let $w = (b \times id) \circ u$. The following commutative diagram of compact polyhedra and PL maps is generated by a one-way chain of four maps, hence can be triangulated by simplicial complexes and simplicial maps.

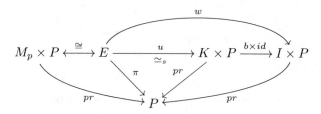

Fix such a triangulation, and let $Z \subset I \times P$ consist of the simplices in $I \times P$ that map isomorphically to their images under $pr \colon I \times P \to P$. This is the part of $I \times P$ over which $u$ and $b \times id$ may have complicated behavior in the $I$-direction, and which we will avoid. Clearly $Z$ meets $I \times p$ in finitely many points, so we can choose a number $t \in (0,1)$ such that $(t,p)$ does not lie in $Z$. Moreover, we can choose a closed neighborhood $B \subset I$ of $t$ and a polyhedral neighborhood $Q \subset P$ of $p$, such that $B \times Q$ is contained in the open star neighborhood of $(t,p)$ within $I \times P$. In particular, $B \times Q$ does not meet $Z$.

For an illustration, see the lower part of Figure 3. The arrows labeled $B$ and $I$ indicate the bicollar $t \in B \subset I$, while the arrows labeled $P$ and $Q$ indicate the polyhedral neighborhood $p \in Q \subset P$. Only part of the subcomplex $Z$ is indicated.

We now focus on the subfamily $E_Q = \pi^{-1}(Q) \to Q$ of thickenings $u \colon M_q \to K$ for $q \in Q$. For each $q \in Q$, let

$$N_q := M_q \times_K (K_0 \times t) = w^{-1}(t,q)$$

be the fiber product. This is obviously a collection of compact polyhedra, and we shall prove that it is a stably framed family ($=$ PL bundle) of $(n-1)$-manifold thickenings of $K_0 \times t$, which are bicollared within the stably framed $n$-manifold thickenings $M_q$ of $K$.

We first claim that we can extend the inclusion

$$F := \bigsqcup_{q \in Q} N_q \subset w^{-1}(B \times Q) =: G$$

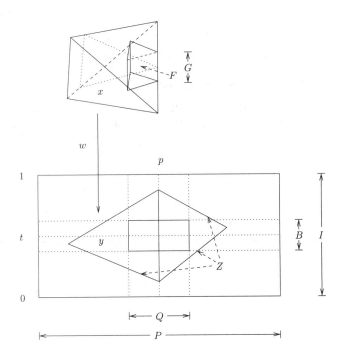

FIGURE 3. The $Q$-parametrized bicollarings $t \in B \subset I$ and $F \subset$
$G \subset E_Q$, where they meet simplices $y \subset I \times P$ and $x \subset E$, with
$w(x) = y$

to an isomorphism

$$F \times B = \bigsqcup_{q \in Q} N_q \times B \xrightarrow{\cong} G$$

over $B \times Q$, that is linear within each simplex of $E$, so as to make $G$ a bicollar of
$F$ in $E_Q$. This isomorphism will be constructed by induction over the simplices
$x$ of $E$, in some order of increasing dimension. See the upper part of Figure 3,
where only the parts of $F$ and $G$ that meet $x$ are indicated. Only the $x$ that
meet $G$ are relevant, and these are mapped by $w$ onto simplices $y = w(x)$ of
$I \times P$ that meet $B \times Q$. Each such $y$ contains $B \times p$, so there is a preferred
linear isomorphism

$$(y \cap (t \times Q)) \times B \cong y \cap (B \times Q)$$

over $B \times Q$. We lift this isomorphism to a linear isomorphism

$$(x \cap F) \times B \cong x \cap G$$

over $B \times Q$, by viewing $x$ as the join of those of its faces that are preimages of
the vertices of $y$, and asking that the isomorphism takes $(f, s)$ with $f \in x \cap F$

and $s \in B$ to $g \in x \cap G$, where $f$ and $g$ have the same join coordinates in each of these (vertex preimage) faces. We also ask that $(f, s)$ and $g$ have the same image in $y \cap (B \times Q)$. This determines an extension of the previously specified isomorphism to $\partial x \cap G$, and proves the claim.

For each $q \in Q$, we now have a PL embedding $e_q \colon N_q \times B \to M_q$ that identifies $N_q \times t$ with $N_q$. For each point $z \in N_q$, the link of $z$ in $N_q$ is PL homeomorphic to the link of the edge $z \times B$ in $M_q$. The latter link is PL homeomorphic to an $(n-2)$-sphere or an $(n-2)$-ball, because $M_q$ is a PL $n$-manifold, hence each $N_q$ is in fact a PL $(n-1)$-manifold, with $\partial N_q = N_q \cap \partial M_q$.

The embedding $e_q$ exhibits a PL bicollaring of $N_q$ in $M_q$. We get a preferred isomorphism $\tau_{N_q} \oplus \epsilon^1 \cong \tau_{M_q} | N_q$ of PL microbundles, so the given stable framing of $M_q$ determines a unique stable framing of $N_q$. Let $v \colon N_q \to K_0$ be the restriction of $u \colon M_q \to K$ to the preimage of $K_0 \times t \cong K_0$. The point inverses of $v$ also occur as point inverses of $u$, hence $v$ is an unthickening map for each $q \in Q$.

It remains to check that the thickenings $v \colon N_q \to K_0$ form a PL locally trivial family over $Q$, and similarly for the bicollars. We restrict the chosen simplicial isomorphism $E \cong M_p \times P$ to $Q \subset P$, and consider the composite embedding

$$e \colon F \times B \xrightarrow{\cong} G \subset E_Q \cong M_p \times Q$$

and the product embedding

$$e_p \times id \colon N_p \times B \times Q \to M_p \times Q,$$

both over $Q$. These agree over $q = p$, so by restricting to a smaller open neighborhood $B' \subset B$ of $t$ and possibly shrinking $Q$ around $p$, we can assume that $e$ factors through an embedding

$$e' \colon F \times B' \to N_p \times B \times Q$$

over $Q$. By construction, this map is linear within each simplex of $E$, and is the standard inclusion $N_p \times B' \subset N_p \times B \times p$ at $q = p$. Hence, for all $q$ sufficiently near $p$ the embedding

$$e'_q \colon N_q \times B' \to N_p \times B$$

contains $N_p \times t$ in its image and, furthermore, for each $z \in N_q$ the composite

$$z \times B' \subset N_q \times B' \xrightarrow{e'_q} N_p \times B \xrightarrow{pr} B$$

is a strictly increasing linear map with $t$ in its (open) image.

After shrinking $Q$ to a smaller polyhedral neighborhood we therefore have, for each $q \in Q$, a unique PL function $s_q \colon N_q \to B'$ such that $(z, s_q(z))$ maps to $t$ by the previous composite. Then the graph $\Gamma_q \subset N_q \times B'$ of $s_q$ maps PL isomorphically to $N_p \times t$ by $e'_q$, and we obtain the desired PL trivialization

$$F = \bigsqcup_{q \in Q} N_q \xleftarrow[\cong]{pr} \bigsqcup_{q \in Q} \Gamma_q \xrightarrow[\cong]{e'} N_p \times t \times Q$$

of the $(n-1)$-manifold family. Translating the graphs a little in the bicollar direction, we also obtain a PL trivialization

$$e'' \colon F \times B'' \xrightarrow{\cong} N_p \times B'' \times Q$$

of the bicollar family, for some closed neighborhood $B'' \subset B'$ of $t$.

Repeating the argument for each point $p \in P$, we get an open covering of $P$ by the interiors of the respective neighborhoods $Q = Q_p$, with the desired transversality property at $t = t_p$. By compactness we find finitely many points $p_i$, with corresponding neighborhoods $P_i = Q_{p_i}$ and bicollar coordinates $t_i = t_{p_i}$, having the properties asserted in the lemma.    $\square$

We now turn to the absolute case of the inductive proof of Theorem 4.3.5, and assume to be in the situation asserted in the lemma. Since we could replace $P_i$ and $P_j$ by $P_i \cup P_j$ if $t_i = t_j$, we may assume that $t_i \neq t_j$ for $i \neq j$. By renumbering the $P_i$ and $t_i$, we may as well assume that $t_i < t_j$ for $i < j$. The proof then proceeds in three steps.

(1) By PL transversality, each $P_i$ parametrizes a family of stably framed $(n-1)$-manifold thickenings of the polyhedron $K_0 \times t_i$, given by the pullback (= fiber product)

$$(4.3.12) \qquad\qquad p \mapsto M_p \times_K (K_0 \times t_i)$$

for $p \in P_i$. By the inductive hypothesis ($\dim(K_0) = k-1$ is less than $\dim(K) = k$, and $\dim(P) \leq n - 2k - 6$ implies $\dim(P_i) \leq (n-1) - 2(k-1) - 6$), each of these can be deformed to a constant family, say

$$p \mapsto N \times t_i,$$

for a fixed $(n-1)$-manifold thickening $N$ of $K_0$. By considering the combined family parametrized by $\coprod_i P_i$, we may assume that this thickening $v \colon N \to K_0$ is the same for each $i$, so that the constant families are "the same."

Also by PL transversality, for each $i$ and small $\epsilon > 0$ the fiber product family

$$(4.3.13) \qquad\qquad p \mapsto M_p \times_K (K_0 \times [t_i - \epsilon, t_i + \epsilon])$$

over $P_i$ is PL isomorphic to the product of the family (4.3.12) with the interval $[t_i - \epsilon, t_i + \epsilon]$. To be precise, this is an isomorphism of $n$-manifold bundles, and commutes with the projection to that bicollar interval, but the two families of unthickening maps to $K_0$ will usually only agree over the part $t_i \times P_i$ of $[t_i - \epsilon, t_i + \epsilon] \times P_i$.

When multiplied with the interval $[t_i - \epsilon, t_i + \epsilon]$, the straightening of the family (4.3.12) to the constant family at $N$ gives a straightening of the product family to the constant family at $N \times [t_i - \epsilon, t_i + \epsilon]$. The PL isomorphism just mentioned then gives a deformation of the pullback family (4.3.13), to the same constant family of manifolds.

This is not quite a straightening, because the unthickening maps $N \to K_0$ at the end of the deformation are only known to be equal to $v$ over the part $t_i \times P_i$. We overcome this defect by a second deformation, where the unthickening map is made independent of the bicollar coordinate in $[t_i - \epsilon, t_i + \epsilon]$, over a gradually increasing neighborhood of $t_i$.

The combined deformation is a straightening of the pullback family (4.3.13) over $P_i$ to the constant family at

$$(4.3.14) \qquad v \times id \colon N \times [t_i - \epsilon, t_i + \epsilon] \to K_0 \times [t_i - \epsilon, t_i + \epsilon].$$

By tapering off the deformation near the boundary of $[t_i - \epsilon, t_i + \epsilon] \times P_i$ (i.e., going through with less and less of the deformation as we get close to the boundary), and making $P_i$ and $\epsilon$ a little smaller, we can arrange that the straightening extends to a deformation of the whole $n$-manifold family $p \mapsto M_p$, parametrized by $P$. The extended deformation can have support over a small neighborhood of the new $[t_i - \epsilon, t_i + \epsilon] \times P_i$, meaning that the deformation is constant away from that neighborhood. In particular, for small $\epsilon$ the deformation does not alter the family over $[t_j - \epsilon, t_j + \epsilon] \times P_j$, for $j \neq i$. We can also retain the property that the interiors of the $P_i$ cover $P$.

Repeating this procedure for each $i$, we get a deformation of the given family of $n$-manifold thickenings $u \colon M_p \to K$, parametrized by $P$, into a new family of the same kind. The deformed family has the additional property that its restriction to $P_i \subset P$ agrees over $K_0 \times [t_i - \epsilon, t_i + \epsilon] \subset K$ with the constant family at (4.3.14), for each $i$.

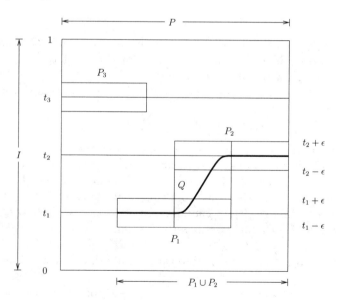

FIGURE 4. Patches of transversality in $I \times P$

(2) Next, we need a construction to reduce the number of patches. If there are two or more patches, then we shall show how the first two patches $P_1$ and $P_2$ may be combined to a single one. By repeating this argument finitely often, we reach the situation with only one patch.

The overlap $P_1 \cap P_2$ parametrizes a family of $n$-manifold thickenings

$$p \mapsto M_p \times_K (K_0 \times [t_1, t_2])$$

for $p \in P_1 \cap P_2$, which sits between the two constant families over $K_0 \times t_1$ and $K_0 \times t_2$, respectively. Identifying $K_0 \times [t_1, t_2]$ with the PL mapping cylinder of the projection

$$K_0 \times \{t_1, t_2\} \to K_0$$

we can put ourselves in the relative situation, by regarding the "between"-family as a family of $n$-manifold thickenings over $K_0 \times \{t_1, t_2\} \to K_0$. By the inductive hypothesis in its relative form, we conclude that we can find a deformation of this family to any constant family. In particular, there is the constant family

$$p \mapsto N \times [t_1, t_2]$$

for $p \in P_1 \cap P_2$, whose manifolds contain the interval $[t_1, t_2]$ as a trivial factor. Thus we can in fact deform into this particular family, relative to the constant family of thickenings of $K_0 \times \{t_1, t_2\}$.

This straightening parametrized by $P_1 \cap P_2$ extends to a deformation of the family

$$p \mapsto M_p \times_K (K_0 \times [t_1, t_2])$$

parametrized by $P$, with support over $K_0 \times [t_1 + \epsilon, t_2 - \epsilon]$ for some small $\epsilon > 0$. This deformation, in turn, extends to a deformation of the original family $p \mapsto M_p$ for $p \in P$, with the same support.

The result of the deformation so far contains the constant family at $v \colon N \to K_0$ over

$$Q = t_1 \times P_1 \cup ([t_1, t_2] \times P_1 \cap P_2) \cup t_2 \times P_2$$

contained in $[0, 1] \times P$. In fact, it is constant over the whole $\epsilon$-neighborhood of $Q$ in the bicollar direction. By possibly shrinking the patches $P_1$ and $P_2$ a little, while maintaining the property that the interiors of all the patches $P_i$ cover $P$, we may ensure that the projection map $pr \colon Q \to P_1 \cup P_2$ (forgetting the bicollar coordinate in $[t_1, t_2]$) admits a PL section $s \colon P_1 \cup P_2 \to Q$. Now the deformed $n$-manifold family contains the bicollared constant $(n-1)$-manifold family over $s(P_1 \cup P_2) \subset Q \subset [t_1, t_2] \times P$, and we just need to move the image of this section up to the slice $t_2 \times P$ to get a family that is transverse to $K_0 \times t_2$ over $P_1 \cup P_2$.

This we achieve by choosing a PL family $p \mapsto h_p$ of self-homeomorphisms of $[0, 1]$, that take the bicollar coordinate of $s(p)$ to $t_2$ for every $p \in P_1 \cup P_2$, and that have support in $[t_1 - \epsilon, t_2 + \epsilon]$ for all $p \in P$. The induced automorphisms $id \times h_p$ of $K_0 \times [0, 1]$ extend to automorphisms of $K$, with the same support. We then alter the deformed family of thickenings by composing the unthickening maps with these (simple) automorphisms. The modified family of thickenings now has the property that the thickening parametrized by $p$ is transverse to $K_0 \times t_2 \subset K$, for each $p \in P_1 \cup P_2$. An isotopy from the identity to the chosen family of self-homeomorphisms then provides the deformation from the original family to the modified family of thickenings.

(3) When there is just one patch, $P = P_1$, we can finish at once. For we have deformed the family of stably framed thickenings $M_p \times_K (K_0 \times t_1)$ to a constant family, and we have extended this straightening to a deformation of the original family of thickenings. We can now split at the pre-image of $K_0 \times t_1$. This leaves us with two families of thickenings, of the PL mapping cylinders

$$K_1 \cup_{K_0} K_0 \times [0, t_1] \qquad \text{and} \qquad K_0 \times [t_1, 1] \cup_{K_0} K_2\,,$$

respectively. The inductive hypothesis, in its relative form, applies to these two families of thickenings. Hence we can find a straightening of each of the two families, relative to the same thickening of $K_0 \times t_1$. These two straightenings are constant near the pre-image of $K_0 \times t_1$, so they can be combined to a straightening of the original family.

### 4.3.15. Inductive step, the relative case.

It remains to discuss the modifications in the general case when $\dim(K) > 0$ and $L$ is not empty. As in the absolute case, we inductively assume that the reformulation of Theorem 4.3.5 holds for thickenings over $L^* \to K^*$ where $\dim(K^*) < \dim(K)$.

First, as pointed out before, the decomposition

$$K = K_1 \cup_{K_0} (K_0 \times [0, 1]) \cup_{K_0} K_2$$

can be obtained from a triangulation of $K$. If $K$ and $L$ have been triangulated so that the map $f\colon L \to K$ is simplicial, then there is a corresponding decomposition

$$L = L_1 \cup_{L_0} (L_0 \times [0, 1]) \cup_{L_0} L_2\,,$$

the map $L \to K$ respects this decomposition, and the map $L_0 \times [0, 1] \to K_0 \times [0, 1]$ is the product of a map $L_0 \to K_0$ with the identity map on $[0, 1]$. So we may assume all this.

Throughout the discussion, manifolds are not over $K$ now, but over the chosen PL mapping cylinder $W_f = W(L \to K)$. This inherits a decomposition

$$W_f = W_1 \cup_{W_0} (W_0 \times [0, 1]) \cup_{W_0} W_2\,,$$

where $W_i = W(L_i \to K_i)$ for $i = 0, 1, 2$. The following diagram may be helpful, where $L_1 = W(L_0 \to L')$ and $L_2 = W(L_0 \to L'')$, etc.

$$
\begin{array}{ccccc}
L' \xleftarrow{L_1} L_0 & \longrightarrow & L_0 \xrightarrow{L_2} L'' \\
\downarrow \quad W_1 \quad \downarrow W_0 & [0,1] & \downarrow W_0 \quad W_2 \quad \downarrow \\
K' \xleftarrow{K_1} K_0 & \longrightarrow & K_0 \xrightarrow{K_2} K''
\end{array}
$$

See also Figure 5.

By the relative analog of Lemma 4.3.11, one finds finitely many patches $P_i$ and levels $t_i \in (0,1)$ so that $u \colon M_p \to W_f$ is PL transverse to $W_0 \times t_i$, for $p \in P_i$. The interiors of the $P_i$ cover $P$, as before. Then $M_p \to W_f$ is PL transverse to $L_0 \times t_i$ for $p \in P_i$, and to $L$ for all $p \in P$, by the assumed local product structure of $M_p$ near $N$.

In the special case just treated, the argument proceeded in three steps:

(1) Straightening of the auxiliary families

$$p \mapsto M_p \times_K (K_0 \times t_i)$$

for $p \in P_i$.

(2) Straightening over the overlap $P_1 \cap P_2$ of

$$p \mapsto M_p \times_K (K_0 \times [t_1, t_2])$$

relative to $K_0 \times \{t_1, t_2\}$.

(3) Straightening over

$$K_1 \cup_{K_0} K_0 \times [0, t_1]$$

(resp. $K_0 \times [t_1, 1] \cup_{K_0} K_2$) relative to $K_0 \times t_1$.

Each of these steps was possible because of the inductive hypothesis applied to a certain subpolyhedron of $K$, or to a map of such subpolyhedra.

In the general case we proceed according to the same pattern. The three steps are:

(1′) Straightening of

$$p \mapsto M_p \times_{W_f} (W_0 \times t_i)$$

relative to

$$p \mapsto M_p \times_{W_f} (L_0 \times t_i),$$

for $p \in P_i$.

(2′) Straightening of

$$p \mapsto M_p \times_{W_f} (W_0 \times [t_1, t_2])$$

relative to

$$p \mapsto M_p \times_{W_f} (L_0 \times [t_1, t_2] \cup W_0 \times \{t_1, t_2\}),$$

for $p \in P_1 \cap P_2$.

(3′) Straightening over

$$W_1 \cup_{W_0} W_0 \times [0, t_1]$$

(resp. $W_0 \times [t_1, 1] \cup_{W_0} W_2$) relative to

$$L_1 \cup_{L_0} (L_0 \times [0, t_1]) \cup_{L_0} W_0 \times t_1$$

(resp. $W_0 \times t_1 \cup_{L_0} (L_0 \times [t_1, 1]) \cup_{L_0} L_2$).

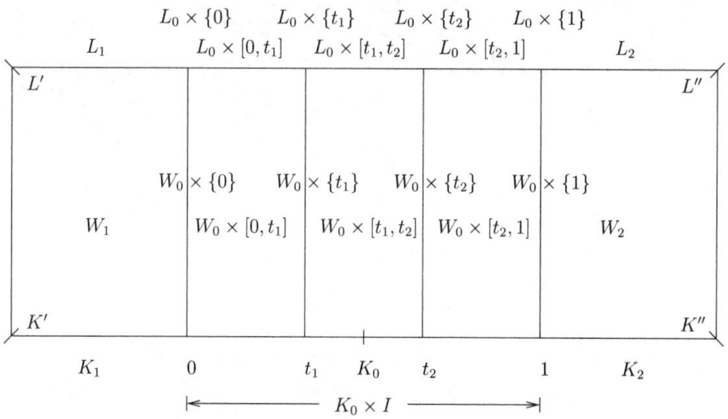

FIGURE 5. Decomposition of $W_f = W(L \to K)$ in terms of smaller PL mapping cylinders

To justify the applicability of the inductive hypothesis, we must know that certain polyhedra can be identified to PL mapping cylinders in a suitable way. This is trivially true in case $(1')$. In case $(2')$ we can identify

$$W_0 \times [t_1, t_2]$$

with the PL mapping cylinder of

$$L_0 \times [t_1, t_2] \cup W_0 \times \{t_1, t_2\} \to K_0$$

(think of a square as the cone on three of its edges, with vertex in the interior of the fourth edge). In a similar way, we can in case $(3')$ identify

$$W_1 \cup_{W_0} W_0 \times [0, t_1]$$

with the PL mapping cylinder of a map

$$L_1 \cup_{L_0} (L_0 \times [0, t_1]) \cup_{L_0} W_0 \to K',$$

where $K_1 = W(K_0 \to K')$, and likewise for $W_0 \times [t_1, 1] \cup_{W_0} W_2$.

This completes the inductive argument, and thus concludes the proof of Theorems 4.3.1 and 4.3.5. $\square$

# Bibliography

[Ak72]    Ethan Akin, *Transverse cellular mappings of polyhedra*, Trans. Amer. Math. Soc. **169** (1972), 401–438.

[Ar70]    Mark Anthony Armstrong, *Collars and concordances of topological manifolds*, Comment. Math. Helv. **45** (1970), 119–128.

[Ba56]    Michael G. Barratt, *Simplicial and semisimplicial complexes*, unpublished manuscript, Princeton University (1956).

[Bo74]    Armand Borel, *Stable real cohomology of arithmetic groups*, Ann. Sci. École Norm. Sup. (4) **7** (1974), 235–272.

[Bo68]    Karol Borsuk, *Concerning homotopy properties of compacta*, Fund. Math. **62** (1968), 223–254.

[Br60]    Morton Brown, *A proof of the generalized Schoenflies theorem*, Bull. Amer. Math. Soc. **66** (1960), 74–76.

[BL74]    Dan Burghelea and Richard Lashof, *The homotopy type of the space of diffeomorphisms. I, II*, Trans. Amer. Math. Soc. **196** (1974), 1–36, 37–50.

[BL77]    Dan Burghelea and Richard Lashof, *Stability of concordances and the suspension homomorphism*, Ann. of Math. (2) **105** (1977), 449–472.

[BLR75]   Dan Burghelea, Richard Lashof and Melvin Rothenberg, *Groups of automorphisms of manifolds. With an appendix ("The topological category") by Erik Kjær Pedersen*, Lecture Notes in Mathematics, vol. 473, Springer–Verlag, Berlin–New York, 1975.

[Ch87]    Thomas A. Chapman, *Piecewise linear fibrations*, Pacific J. Math. **128** (1987), 223–250.

[Cl81]    Mónica Clapp, *Duality and transfer for parametrized spectra*, Arch. Math. (Basel) **37** (1981), 462–472.

[Co67]    Marshall M. Cohen, *Simplicial structures and transverse cellularity*, Ann. of Math. (2) **85** (1967), 218–245.

[Co70]    Marshall M. Cohen, *Homeomorphisms between homotopy manifolds and their resolutions*, Invent. Math. **10** (1970), 239–250.

[Co73]    Marshall M. Cohen, *A course in simple-homotopy theory*, Graduate Texts in Mathematics, vol. 10, Springer–Verlag, New York, 1973.

[Du97]    Bjørn Ian Dundas, *Relative K-theory and topological cyclic homology*, Acta Math. **179** (1997), 223–242.

[EZ50]    Samuel Eilenberg and Joseph A. Zilber, *Semi-simplicial complexes and singular homology*, Ann. of Math. (2) **51** (1950), 499–513.

[FH78]    F. Thomas Farrell and Wu-Chung Hsiang, *On the rational homotopy groups of the diffeomorphism groups of discs, spheres and aspherical manifolds*, Algebraic and geometric topology (Stanford Univ., Stanford, CA, 1976), Part 1, Proc. Sympos. Pure Math., vol. XXXII, Amer. Math. Soc., Providence, RI, 1978, pp. 325–337.

[FP90]    Rudolf Fritsch and Renzo A. Piccinini, *Cellular Structures in Topology*, Cambridge Studies in Advanced Mathematics, vol. 19, Cambridge University Press, Cambridge, 1990.

[FP67]    Rudolf Fritsch and Dieter Puppe, *Die Homöomorphie der geometrischen Realisierungen einer semisimplizialen Menge und ihrer Normalunterteilung*, Arch. Math. (Basel) **18** (1967), 508–512.

[GZ67]    Peter Gabriel and Michel Zisman, *Calculus of fractions and homotopy theory*, Ergebnisse der Mathematik und ihrer Grenzgebiete, vol. 35, Springer–Verlag, Berlin–Heidelberg–New York, 1967.

[GH99]    Thomas Geisser and Lars Hesselholt, *Topological cyclic homology of schemes*, Algebraic $K$-theory (Seattle, WA, 1997), Proc. Sympos. Pure Math., vol. 67, Amer. Math. Soc., Providence, RI, 1999, pp. 41–87.

[GJ99]    Paul G. Goerss and John F. Jardine, *Simplicial homotopy theory*, Progress in Mathematics, vol. 174, Birkhäuser, Basel, 1999.

[Go90a]   Thomas G. Goodwillie, *A multiple disjunction lemma for smooth concordance embeddings*, Mem. Amer. Math. Soc. **86** (1990), no. 431.

[Go90b]   Thomas G. Goodwillie, *Calculus. I. The first derivative of pseudoisotopy theory*, $K$-Theory **4** (1990), 1–27.

[HP64]    André Haefliger and Valentin Poenaru, *La classification des immersions combinatoires*, Inst. Hautes Études Sci. Publ. Math. **23** (1964), 75–91.

[HW65]    André Haefliger and C. Terence C. Wall, *Piecewise linear bundles in the stable range*, Topology **4** (1965), 209–214.

[Ha75]    Allen E. Hatcher, *Higher simple homotopy theory*, Ann. of Math. (2) **102** (1975), 101–137.

[Ha76]    Allen E. Hatcher, *Homeomorphisms of sufficiently large $P^2$-irreducible 3-manifolds*, Topology **15** (1976), 343–347.

[Ha78]    Allen E. Hatcher, *Concordance spaces, higher simple-homotopy theory, and applications*, Algebraic and geometric topology (Stanford Univ., Stanford, CA, 1976), Part 1, Proc. Sympos. Pure Math., vol. XXXII, Amer. Math. Soc., Providence, RI, 1978, pp. 2–21.

[Ha99]    Allen E. Hatcher, *Spaces of incompressible surfaces*, arXiv:math.GT/9906074.

[Ha02]    Allen E. Hatcher, *Algebraic Topology*, Cambridge University Press, Cambridge, 2002.

[He94]    Lars Hesselholt, *Stable topological cyclic homology is topological Hochschild homology*, $K$-theory (Strasbourg, 1992), Astérisque, vol. 226, 1994, pp. 8–9, 175–192.

[HSS00]   Mark Hovey, Brooke Shipley and Jeff Smith, *Symmetric spectra*, J. Amer. Math. Soc. **13** (2000), 149–208.

[HJ82]    Wu-Chung Hsiang and Bjørn Jahren, *A note on the homotopy groups of the diffeomorphism groups of spherical space forms*, Algebraic $K$-theory, Part II, Lecture Notes in Math., vol. 967, Springer, 1982, pp. 132–145.

[HJ83]    Wu-Chung Hsiang and Bjørn Jahren, *A remark on the isotopy classes of diffeomorphisms of lens spaces*, Pacific J. Math. **109** (1983), 411–423.

[Hu69]    John F. P. Hudson, *Piecewise linear topology*, University of Chicago Lecture Notes, W. A. Benjamin, Inc., New York–Amsterdam, 1969.

[HTW90]   C. B. Hughes, L. R. Taylor and E. B. Williams, *Bundle theories for topological manifolds*, Trans. Amer. Math. Soc. **319** (1990), 1–65.

[HW41]    Witold Hurewicz and Henry Wallman, *Dimension Theory*, Princeton Mathematical Series, vol. 4, Princeton University Press, 1941.

[Ig88]    Kiyoshi Igusa, *The stability theorem for smooth pseudoisotopies*, $K$-Theory **2** (1988), 1–355.

[Jo32]    Ingebrigt Johansson, *Über die Invarianz der topologischen Wechselsumme $a_0 - a_1 + a_2 - a_3 + \cdots$ gegenüber Dimensionsänderungen*, Avh. Norske Vid. Akad. Oslo **1** (1932), 1–8.

[Ka57]    Daniel M. Kan, *On c.s.s. complexes*, Amer. J. Math. **79** (1957), 449–476.

[KS77]    Robion C. Kirby and Laurence C. Siebenmann, *Foundational essays on topological manifolds, smoothings, and triangulations*, Annals of Mathematics Studies, vol. 88, Princeton University Press, 1977.

[Ki64]    James M. Kister, *Microbundles are fibre bundles*, Ann. of Math. (2) **80** (1964), 190–199.

[KL66]    Nicolaas H. Kuiper and Richard K. Lashof, *Microbundles and bundles. I. Elementary theory*, Invent. Math. **1** (1966), 1–17.

[La69a]   R. Chris Lacher, *Cell-like spaces*, Proc. Amer. Math. Soc. **20** (1969), 598–602.

[La69b]   R. Chris Lacher, *Cell-like mappings. I*, Pacific J. Math. **30** (1969), 717–731.

[Li73]    Joachim Lillig, *A union theorem for cofibrations*, Arch. Math. (Basel) **24** (1973), 410–415.

[Ma71]    Saunders Mac Lane, *Categories for the Working Mathematician*, Graduate Texts in Mathematics, vol. 5, Springer, 1971.

[Ma61]    Barry Mazur, *A note on some contractible 4-manifolds*, Ann. of Math. (2) **73** (1961), 221–228.

[Mc64]    Daniel R. McMillan, *A criterion for cellularity in a manifold*, Ann. of Math. (2) **79** (1964), 327–337.

[Me67]    Wolfgang Metzler, *Beispiele zu Unterteilungsfragen bei CW- und Simplizialkomplexen*, Arch. Math. (Basel) **18** (1967), 513–519.

[Mi59]    John W. Milnor, *On spaces having the homotopy type of a CW-complex*, Trans. Amer. Math. Soc. **90** (1959), 272–280.

[Mi64]    John W. Milnor, *Microbundles. I*, Topology **3(1)** (1964), 53–80.

[Mi66]    John W. Milnor, *Whitehead torsion*, Bull. Amer. Math. Soc. **72** (1966), 358–426.

[Po60]    Valentin Poenaru, *Les decompositions de l'hypercube en produit topologique*, Bull. Soc. Math. France **88** (1960), 113–129.

[Qu73]    Daniel Quillen, *Higher algebraic K-theory. I*, Algebraic K-theory, I: Higher K-theories (Proc. Conf., Battelle Memorial Inst., Seattle, WA, 1972), Lecture Notes in Math., vol. 341, Springer, Berlin, 1973, pp. 85–147.

[Ro02]    John Rognes, *Two-primary algebraic K-theory of pointed spaces*, Topology **41** (2002), 873–926.

[Ro03]    John Rognes, *The smooth Whitehead spectrum of a point at odd regular primes*, Geom. Topol. **7** (2003), 155–184.

[RS72]    Colin P. Rourke and Brian J. Sanderson, *Introduction to piecewise-linear topology*, Ergebnisse der Mathematik und ihrer Grenzgebiete, vol. 69, Springer–Verlag, New York–Heidelberg, 1972.

[Se68]    Graeme Segal, *Classifying spaces and spectral sequences*, Inst. Hautes Études Sci. Publ. Math. **34** (1968), 105–112.

[Se74]    Graeme Segal, *Categories and cohomology theories*, Topology **13** (1974), 293–312.

[Si70]    Laurent C. Siebenmann, *Infinite simple homotopy types*, Nederl. Akad. Wetensch. Proc. Ser. A 73 = Indag. Math. **32** (1970), 479–495.

[Si72]    Laurent C. Siebenmann, *Approximating cellular maps by homeomorphisms*, Topology **11** (1972), 271–294.

[St86]    Mark Steinberger, *The classification of PL fibrations*, Michigan Math. J. **33** (1986), 11–26.

[SW84]    Mark Steinberger and James West, *Covering homotopy properties of maps between C.W. complexes or ANRs*, Proc. Amer. Math. Soc. **92** (1984), 573–577.

[St70]    Ralph Stöcker, *Whiteheadgruppe topologischer Räume*, Invent. Math. **9** (1969/1970), 271–278.

[Th79]    Robert W. Thomason, *Homotopy colimits in the category of small categories*, Math. Proc. Cambridge Philos. Soc. **85** (1979), 91–109.

[Ur24]    Paul Urysohn, *Der Hilbertsche Raum als Urbild der metrischen Räume*, Math. Ann. **92** (1924), 302–304.

[Wa78a]   Friedhelm Waldhausen, *Algebraic K-theory of generalized free products*, Ann. of Math. (2) **108** (1978), 135–256.

[Wa78b]   Friedhelm Waldhausen, *Algebraic K-theory of topological spaces. I*, Algebraic and geometric topology (Stanford Univ., Stanford, CA, 1976), Part 1, Proc. Sympos. Pure Math., vol. XXXII, Amer. Math. Soc., Providence, RI, 1978, pp. 35–60.

[Wa82]    Friedhelm Waldhausen, *Algebraic K-theory of spaces, a manifold approach*, Current trends in algebraic topology (London, Ontario, 1981), Part 1, CMS Conf.

Proc., vol. 2, Amer. Math. Soc., 1982, pp. 141–184, available for download at http://www.math.uni-bielefeld.de/~fw/.

[Wa85]    Friedhelm Waldhausen, *Algebraic K-theory of spaces*, Algebraic and geometric topology (New Brunswick, NJ, 1983), Lecture Notes in Math., vol. 1126, Springer, Berlin, 1985, pp. 318–419.

[Wa87a]   Friedhelm Waldhausen, *Algebraic K-theory of spaces, concordance, and stable homotopy theory*, Algebraic topology and algebraic K-theory (Princeton, NJ, 1983), Ann. of Math. Stud., vol. 113, Princeton University Press, 1987, pp. 392–417.

[Wa87b]   Friedhelm Waldhausen, *An outline of how manifolds relate to algebraic K-theory*, Homotopy theory (Durham, 1985), London Math. Soc. Lecture Note Ser., vol. 117, Cambridge University Press, Cambridge, 1987, pp. 239–247.

[Wa96]    Friedhelm Waldhausen, *On the construction of the Kan loop group*, Doc. Math. **1** (1996), 121–126.

[Wa66]    C. Terence C. Wall, *Classification problems in differential topology. IV. Thickenings*, Topology **5** (1966), 73–94.

[Wa70]    C. Terence C. Wall, *Surgery on compact manifolds*, London Mathematical Society Monographs, No. 1, 1970.

[WW88]    Michael Weiss and Bruce Williams, *Automorphisms of manifolds and algebraic K-theory. I*, K-Theory **1** (1988), 575–626.

[WW95]    Michael Weiss and Bruce Williams, *Assembly*, Novikov conjectures, index theorems and rigidity, Vol. 2 (Oberwolfach, 1993), London Math. Soc. Lecture Note Ser., vol. 227, Cambridge University Press, Cambridge, 1995, pp. 332–352.

[WW01]    Michael Weiss and Bruce Williams, *Automorphisms of manifolds*, Surveys on surgery theory, Vol. 2, Ann. of Math. Stud., vol. 149, Princeton University Press, 2001, pp. 165–220.

[Wh39]    John Henry Constantine Whitehead, *Simplicial spaces, nuclei and m-groups*, Proc. Lond. Math. Soc., II. Ser. **45** (1939), 243–327.

[Ze64]    Eric Christopher Zeeman, *Relative simplicial approximation*, Proc. Cambridge Philos. Soc. **60** (1964), 39–43.

# Symbols

$*$, one-point space, 2

$A(X)$, algebraic $K$-theory of spaces, 2, 16
$\mathbf{A}(X)$, algebraic $K$-theory spectrum, 2, 16

$B(X)$, Barratt nerve, 35
$\beta_n^\rho$, pseudo-barycenter (w.r.t. a degeneracy operator), 48

$C_f$, Cohen's PL mapping cylinder, 156
$\mathrm{cone}(X)$, cone, 40
$\mathrm{cone}^{op}(X)$, backward cone, 61
$s\mathcal{C}^h(X)$, category of simplicial sets, 14, 101
$s\widetilde{\mathcal{C}}_\bullet^h(X)$, simplicial category of simplicial sets, 102

$s\mathcal{D}^h(X)$, category of non-singular simplicial sets, 14, 101
$s\mathcal{D}_\bullet^h(X)$, $s\widetilde{\mathcal{D}}_\bullet^h(X)$, simplicial categories of non-singular simplicial sets, 102
$d_X\colon Sd(X) \to X$, last vertex map, 41
$d_X^{op}\colon Sd^{op}(X) \to X$, first vertex map, 42
$\Delta^q$, affine $q$-simplex, 7, 139
$\Delta^q$, simplicial $q$-simplex, 13, 29, 99

$s\mathcal{E}^h(K)$, category of polyhedra, 15, 102
$s\mathcal{E}_\bullet^h(K)$, $s\widetilde{\mathcal{E}}_\bullet^h(K)$, simplicial categories of polyhedra, 10, 103

$H(M) = H^{CAT}(M)$, CAT $h$-cobordism space, 7
$H(M)^c$, collared CAT $h$-cobordism space, 7
$H(M)^f$, space of collared, stably framed $h$-cobordisms, 141
$\mathcal{H}^{CAT}(M)$, stable $h$-cobordism space, 8

$I(X) = B(Sd(X))$, improvement functor, 68
$I$, interval for collaring, 7

$J$, interval for stabilizing, 7

$Lk(p, Y)$, link, 115
$\Lambda_i^n$, horn, 108

$M(f)$, reduced mapping cylinder, 58
$\mathcal{M}_\bullet^n$, space of stably framed $n$-manifolds, 140
$h\mathcal{M}_\bullet^n$, simplicial category of stably framed $n$-manifolds, 140
$\langle \mu \rangle_x^f$, pseudo-barycenter (of a face, w.r.t. a map), 49

$\mathcal{R}_f(X)$, category of finite retractive spaces, 15
$\hat{\rho}$, maximal section, 46

$Sd(X)$, normal subdivision, 37
$Sd^{op}(X)$, op-normal subdivision, 42
$\mathrm{simp}(X)$, simplex category, 37
$\mathrm{simp}_\eta(X)$, augmented simplex category, 40

$S^n(K)$, space of stably framed simple $n$-manifolds, 150

$St(p, Y)$, star, 115

$T(f)$, mapping cylinder, 30, 56

$T(f) \to M(f)$, cylinder reduction map, 59

$T^n(K)$, space of stably framed $n$-manifold thickenings, 153

$T^n(L \to K, N)$, space of relative thickenings, 157

$\tau_M$, tangent microbundle, 139

$W_f$, Whitehead's PL mapping cylinder, 156

$\mathrm{Wh}^{DIFF}(X)$, DIFF Whitehead space, 19

$\mathbf{Wh}^{DIFF}(X)$, DIFF Whitehead spectrum, 19

$\mathrm{Wh}^{PL}(X)$, PL Whitehead space, 18

$\mathbf{Wh}^{PL}(X)$, PL Whitehead spectrum, 18

$\mathrm{Wh}^{TOP}(X)$, TOP Whitehead space, 18

$\mathbf{Wh}^{TOP}(X)$, TOP Whitehead spectrum, 18

$x^{\#}$, non-degenerate part, 35

$\bar{x}$, representing map, 14

$X^{op}$, opposite simplicial set, 42

$f/Y$, left fiber of a functor, 134

$Y/f$, right fiber of a functor, 134

$f/([q], Y)$, left fiber of a simplicial functor, 135

$([q], Y)/f$, right fiber of a simplicial functor, 135

# Index

0-dimensional up to isomorphism, 137

2-out-of-3 property (for weak homotopy equivalences), 30

absolute neighborhood retract (ANR), 32
algebraic $K$-theory of spaces, $A(X)$, 2, 16
approximate lifting property for polyhedra (ALP), 75
assembly map, 17
augmented simplex category, $\text{simp}_\eta(X)$, 40

back inclusion, 59
backward cone, $\text{cone}^{op}(X)$, 61
backward reduced mapping cylinder, 58
Barratt nerve, $B(X)$, 35
barycenter, 36
base gluing lemma for Serre fibrations, 95
base inclusion, 40
block, 130

CAT bundle, 7
CAT bundle relative to a product subbundle, 7
CAT concordance, 12
CAT $h$-cobordism space, $H(M) = H^{CAT}(M)$, 7
CAT pseudo-isotopy, 12
Čech homotopy type (shape), 32
cell-like (space or map), 32
cellular (subspace of a manifold), 32

cofibration (of simplicial sets), 14
cofibrations under $X$, 99
collared CAT $h$-cobordism space, $H(M)^c = H^{CAT}(M)^c$, 7
composition property (for simple maps), 29
concordance, 12
concordance stability theorem, 22, 23, 153
cone, $\text{cone}(X)$, 40
cylinder coordinate projection, 59
cylinder projection, 59
cylinder reduction map, $T(f) \to M(f)$, 59

deformation into a constant family, 159
deformation (of a family of thickenings), 159
degeneracy operator, 34
desingularization, 39
desingularized simplicial sets, 99
DIFF Whitehead space, 19
DIFF Whitehead spectrum, 19
dual-collapsible map, 118

Eilenberg–Zilber lemma, 35
elementary expansion, 109
elementary simplicial expansion, 109
embedded horn, 109
$\epsilon$-lifting, 77
equivalent triangulations, 126
Euclidean neighborhood retract (ENR), 32
Euclidean polyhedra, 99
expansion, 109

extension property, 113

face operator, 34
family of manifolds (parametrized
    by $P$), 139
fiber gluing lemma for Serre fibra-
    tions, 94
fibrant simplicial set, 113
filling a horn, 109
filling all horns, 114
finite cofibration, 14, 105
finite simplicial set, 13
first vertex map, $d_X^{op} \colon Sd^{op}(X) \to$
    $X$, 42
forward reduced mapping cylinder,
    58
front face, 40
front inclusion, 59
functoriality of collared $h$-
    cobordism space, 8

global transversality in patches, 163
gluing lemma (for closed cofibra-
    tions), 138
gluing lemma (for simple maps), 30
gluing lemma (for weak homotopy
    equivalences), 30

$h$-cobordism theorem, 1
hereditary, 29
homotopical dimension, 161
homotopy equivalence over the tar-
    get, 60
horn in a simplicial set, 109
horn, $\Lambda_i^n$, 108

improvement functor, $I$, 68
iterated reduced mapping cylinder,
    64

Kan complex, 113

last vertex map, $d_X \colon Sd(X) \to X$,
    41

left fiber (of a simplicial functor),
    $f/([q], Y)$, 135
left fiber (of a functor), $f/Y$, 134
left ideal, 65
linear subdivision, 126
link, $Lk(p, Y)$, 115
locally trivial up to simple maps, 78

mapping cylinder, $T(f)$, 30, 56
maximal section, $\hat{\rho}$, 46
mesh (of a triangulation), 77

nerve (of a small category), 34
$n$-manifold thickening, 153
non-degenerate part, $x^{\#}$, 35
non-singular simplicial set, 14
normal subdivision, $Sd(X)$, 37

one-way tree, 130
op-normal subdivision, $Sd^{op}(X)$, 42
opposite simplicial set, $X^{op}$, 42
op-regular (simplex), 70
op-regular (simplicial set), 70
op-regularization, 70

PL bundle, 99, 139
PL map, 126
PL mapping cylinder (Cohen's,
    $C_f$), 156
PL mapping cylinder (Whitehead's,
    $W_f$), 156
PL Serre fibration, 10, 100
PL structure, 126
PL Whitehead space, 18
PL Whitehead spectrum, 18
polyhedral realization, 14, 99, 127
polyhedron, 126
preimage of the barycenter (of a
    face), 84
principle of global transversality in
    patches, 163
proper degeneracy operator, 34
proper face operator, 34
proper homotopy equivalence, 32

property $UV^\infty$, 32

pseudo-barycenter (of a face, w.r.t. a map), 48

pseudo-barycenter (w.r.t. a degeneracy operator), 48

pseudo-isotopy, 12

pullback map (of $h$-cobordism spaces with retractions), 11

pullback map (of preimages), 84

pullback property (for simple maps), 29

Quillen's Theorem A, 134

Quillen's Theorem B, 135

realization lemma, 138

reduced cylinder projection, 59

reduced mapping cylinder, $M(f)$, 58

regular (CW-complex), 69

regular (simplicial set), 69

relative approximate lifting property (relative ALP), 75

relative mapping cylinder, 115

representing map, $\bar{x}$, 14

right cancellation property (for simple maps), 29

right fiber (of a functor), $Y/f$, 134

right fiber (of a simplicial functor, $([q], Y)/f$), 135

$s$-cobordism theorem, 2

Serre fibration (of simplicial sets), 80, 100

simple cylinder reduction, 60

simple map (of compact polyhedra), 9

simple map (of finite simplicial sets), 13, 29

simple map (of topological spaces), 76

simple $n$-manifold over a polyhedron, 150

simple onto its image, 67

simple over $V$, 34

simplex category, $\mathrm{simp}(X)$, 37

simplex-wise Serre fibration, 94

simplicial expansion, 109

simplicial set, 29

simplicial homotopy equivalence over the target, 60

simplicially collapsible, 91

space of collared, stably framed $h$-cobordisms, $H(M)^f$, 141

space of stably framed $n$-manifolds, $\mathcal{M}_\bullet^n$, 140

sphere spectrum, 16

stabilization map (of category of polyhedra), 10

stabilization map (of $h$-cobordisms), 8

stabilization (of a homotopy functor), 19

stable framing, 140

stable $h$-cobordism space, $\mathcal{H}^{CAT}(M)$, 8

stable parametrized $h$-cobordism theorem, 1

stably framed family of manifolds, 140

stably framed map, 140

stably framed $n$-manifold thickening, 153

star, $St(p, Y)$, 115

starring a convex block, 130

starring an ordered simplicial complex, 115

stellar subdivision, 115

straightening, 159

support of a point, 131

tangent microbundle, $\tau_M$, 139

$(T, \epsilon)$-lifting, 77

Theorem A$'$, 136

Theorem B$'$, 136

thickening of a polyhedron, 153

thickening over $L \to K$ relative to $N$, 157

TOP Whitehead space, 18

TOP Whitehead spectrum, 18

trace of a stellar subdivision, 115
transition map (of left fibers), 135
transition map (of right fibers), 106,
    134
transition map of the first kind, 136
transition map of the second kind,
    136

triangulation, 126
trivial up to a simple map, 78

unnatural homeomorphisms, 36
unthickening map, 153

weak homotopy equivalence, 14